Urban Agriculture

Series editors
Christine Aubry, AgroParisTech, INRA UMR SADAPT, Paris, France
Éric Duchemin, Université du Québec à Montréal Institut des Science de Environment, Montreal, Québec, Canada
Joe Nasr, Centre for Studies in Food Security, Ryerson University, Toronto, Ontario, Canada

This new *Urban Agriculture* Book Series at Springer is for researchers, professionals, policy-makers and practitioners working on agriculture in and near urban areas. Urban agriculture (UA) can serve as a multifunctional resource for resilient food systems and socio-culturally, economically and ecologically sustainable cities.

For the Book Series Editors, the main objective of this series is to mobilize and enhance capacities to share UA experiences and research results, compare methodologies and tools, identify technological obstacles, and adapt solutions. By diffusing this knowledge, the aim is to contribute to building the capacity of policy-makers, professionals and practitioners in governments, international agencies, civil society, the private sector as well as academia, to effectively incorporate UA in their field of interests. It is also to constitute a global research community to debate the lessons from UA initiatives, to compare approaches, and to supply tools for aiding in the conception and evaluation of various strategies of UA development.

The concerned scientific field of this series is large because UA combines agricultural issues with those related to city management and development. Thus, we propose an interdisciplinary Book Series where environmental sciences, agronomy, urban and regional planning, architecture, landscape design, economics, social sciences, soil sciences, public health and nutrition come together, recognizing UA's contribution to meeting society's basic needs, feeding people, structuring the cities while shaping their development. All these scientific fields are of interest for this Book Series. Books in this Series will analyze UA research and actions; program implementation, urban policies, technological innovations, social and economic development, management of resources (soil/land, water, wastes …) for or by urban agriculture, are all pertinent here.

This Book Series will include a mix of edited, coauthored, and single-authored books. These books could be based on research programs, conference papers, or other collective efforts, as well as completed theses or entirely new manuscripts.

More information about this series at http://www.springer.com/series/11815

Christophe-Toussaint Soulard · Coline Perrin
Elodie Valette
Editors

Toward Sustainable Relations Between Agriculture and the City

Springer

Editors
Christophe-Toussaint Soulard
Innovation, INRA, CIRAD,
Montpellier Supagro, University of Montpellier
Montpellier, France

Coline Perrin
Innovation, INRA, CIRAD,
Montpellier Supagro, University of Montpellier
Montpellier, France

Elodie Valette
ART-DEV
University of Montpellier, CIRAD,
CNRS, University of Paul Valery,
University of Perpignan
Montpellier, France

ISSN 2197-1730 ISSN 2197-1749 (electronic)
Urban Agriculture
ISBN 978-3-319-71035-8 ISBN 978-3-319-71037-2 (eBook)
https://doi.org/10.1007/978-3-319-71037-2

Library of Congress Control Number: 2017963134

Printed on acid-free paper

This Springer imprint is published by Springer Nature
The registered company is Springer International Publishing AG
The registered company address is: Gewerbestrasse 11, 6330 Cham, Switzerland

Contents

List of Figures

List of Tables

Chapter 1
Relations Between Agriculture and the City in Europe and the Mediterranean

Christophe-Toussaint Soulard, Coline Perrin, and Elodie Valette

Abstract The book gives an overview of frameworks, methods, and case studies for the analysis of the relations between agriculture and the city, in Europe and the Mediterranean. Its origin is the DAUME Project (Sustainability of urban agriculture in the Mediterranean) and the 5th annual Conference of the Sustainable Food Planning group of AESOP, held in Montpellier (France) in 2013. The book provides a set of approaches of the sustainability of urban food systems from an actors' perspective. The Part I presents systemic approaches of agricultural-urban interactions at the city-region scales in France, Egypt, Italy and Morocco. Local food issues, agriculture-urban relations, short food chains and urban livestock are taken as examples to develop systemic approaches showing both integrative and dualism processes linking agriculture and the city. The Part II deals with methods and tools for urban planning and local development, in order to design and assess sustainable food systems. At the city-region scale, chapters show how to estimate relevant boundaries of a sustainable foodshed, to design tools including local food supply In urban planning, and to evaluate contributions of local projects to global sustainability. The Part III inventories the recent changes in urban agriculture and the new forms of governance which are emerging in European cities (Athens, Berlin, Lisboa, Montpellier, Paris and Zurich). Referring to urban agriculture, chapters show how sustainable pathways can be fostered by a wide range of multiscale grassroots initiatives (farms, gardens, buildings, urban green areas …) embedded in transitioning trends of sustainable development.

C.-T. Soulard (✉) · C. Perrin
Innovation, INRA, CIRAD, Montpellier Supagro, University of Montpellier, Montpellier, France
e-mail: christophe.soulard@inra.fr; coline.perrin@inra.fr

E. Valette
ART-DEV, University of Montpellier, CIRAD, CNRS, University of Paul Valery, University of Perpignan, Montpellier, France
e-mail: elodie.valette@cirad.fr

C.-T. Soulard et al. (eds.), *Toward Sustainable Relations Between Agriculture and the City*, Urban Agriculture, https://doi.org/10.1007/978-3-319-71037-2_1

Keywords Urban agriculture · Food system · Sustainability · Governance

1.1 Introduction and Conceptual Framework

The decision to write and publish this book on relations between agriculture and the city (i.e., agricultural-urban relationships) stems from the DAUME Project (Sustainability of urban agriculture in the Mediterranean), led by C.T. Soulard, E. Valette and C. Perrin, from 2011 to 2015.[1] One of the main results of this project was the observation of large discrepancies in the way agriculture responds to urbanization in five city-regions studied on both sides of the Mediterranean. As shown in Soulard et al. (2017), urbanization tends to accentuate agricultural diversity: according to local contexts, urban and peri-urban agriculture may decline, demonstrate its resilience, or even enter a new phase of growth in response to the urban demand. There are also strong differences in the way urban planners address agricultural and environmental issues. Sustainable development is a challenge for both agri-food sector and urban planning. Systemic and comparative approaches achieved in the DAUME project showed that sustainability is a relative concept whose understanding implies to take into account how actors, activities and policies interact at different scales of time and space. Referring to urban agriculture, it can be fostered by a series of local changes and innovative initiatives embedded in more global transitioning trends of development (Cohen and Ilieva 2015; Soulard et al. 2016).

This book also originates from the 5th annual Conference of the Sustainable Food Planning group of AESOP, held in Montpellier (France) in October 2013.[2] Organized by C. Perrin and C.T. Soulard, the conference focused on innovations in the urban food systems analyzed through flows and the reframing of urban foodsheds, the land and farming for and in the city, and the governance and food justice. This conference showed how the rise of food issues in the cities changes the significations, the functions and the challenges of urban and peri-urban agriculture. This challenge has been known for a long time in the global South, because of food security and safety issues (Hamilton et al. 2013; Moustier 2017). In European and Mediterranean countries, this is a more recent issue. This conference also focused on how food issues renew the question of the role of agriculture in the sustainable development of cities, through "food justice"-driven approaches in the American studies (Gottlieb and Joshi 2010; Heynen et al. 2012) or "food planning"-driven approaches in the European studies (Morgan 2009; Viljoen and Wiskerke 2012). All

[1] DAUME (*Durabilité des agricultures urbaines en Méditerranée*) is a research project funded by the French National Research Agency (ANR-2010-STRA-007). It brought together urban and agricultural scholars from France (Montpellier), Algeria (Constantine), Italy (Pisa), Morocco (Meknès) and Portugal (Lisbon).

[2] AESOP (Association of European Schools of Planning), Sustainable Food Planning group: https://aesopsfp.wordpress.com/

over, researchers highlighted the way local actors elaborate new food policies (Mansfield and Mendes 2013; Sonnino 2014; Moragues-Faus and Morgan 2015).

The original objective of the book is to analyze the sustainable agricultural-urban relations through the actors practices and strategies. Theoretical and methodological approaches may help to achieve a better understanding of these relations: (1) unfolding the complex nature of urban food systems; (2) developing a specific insight on agricultural-urban relations; (3) promoting a actors-based analysis of sustainability.

1.1.1 Thinking "Urban Food Systems" as a Perspective

More than a concept, the "urban food system" refers to approaches considering food issues as an essential component of a sustainable urban way of life. According to Pothukuchi and Kaufman (1999), urban food systems are less visible than other systems in the city such as transportation, housing, employment, or even the environment. A reason for this low visibility is the long-term divide between urban development and food issues (Steel 2008). It comes from an historic process by which issues and policies came to be defined as urban, neglecting agriculture and food issues. In the literature, Chaboud et al. (2013) distinguish three trends in the research works on urban food systems. First, scholars analyze the organizations, flows and spatial dynamics linking urban development and food, especially through the analysis of logistics of food procurement or of urban metabolism. Cities are seen as places that import, transform and export flows of food nutrients. Foodsheds (Horst and Gaolach 2015), foodscapes (Cummins and Macintyre 2002) and food deserts (Walker et al. 2010) are key concepts of these approaches. Secondly, scholars analyze local food systems focusing on the causal relationships between local food production and urban sustainability. More and more urban-dwellers and local authorities engage in alternative food networks through urban gardening, short food chains, direct selling by producers, food education, etc. Third, scholars analyze food planning and public policies. A key issue is the inclusion of agriculture and food in the design of urban planning (Viljoen et al. 2005; Viljoen and Bohn 2014). These three approaches for a "new food equation" (Morgan and Sonnino 2010) demand to elaborate public policies promoting urban food systems locally embedded and open to the global market (i.e., avoiding the "local trap") (Morgan 2010).

In recent years, there have been more and more exchanges between scholars from these various approaches. The conferences of AESOP-SFP are examples of places where an interdisciplinary community of researchers aims to address sustainability together with the issues of food procurement, food movement and food planning. Other international networks, connecting scholars with stakeholders, contribute to this effort, like the "food for the city" initiative of the FAO,[3] which aims at addressing the challenges of urbanization by building more sustainable and

[3] Food for the Cities Initiative: http://www.fao.org/fcit/fcit-home/fr/

resilient food systems, or the RUAF foundation promoting sustainable urban agriculture and food systems all over the world.[4] Moreover, networks emerge among cities willing to engage into urban food strategies: the Milan urban food policy pact[5] was signed in 2015 by 133 cities around the world. In the UK, the Sustainable Food Cities[6] network gathered in 2016 56 cities. In France, 26 cities are members of the Terres en ville[7] network.

All these initiatives and networks aim to advance the sustainability of urban food systems. However, which specific role does agriculture play in sustainability? How to analyze agricultural-urban relations?

1.1.2 Analyzing Agricultural-Urban Relations

The concept of urban food system gives a new perspective to the agricultural studies. First, it is important not to separate the diverse forms of urban agriculture, – social vs. entrepreneurial, individual vs. community (McClintock 2010), from the peri-urban and rural agriculture that is primarily commercial (Bryant and Charvet 2003). It is more relevant to consider all these agricultures from a multifunctional perspective (Aubry et al. 2012; Duchemin et al. 2010), but also through the geographical lens of the agricultural-urban relations, involving environment, land tenure, planning, employment, social, health. The scientific community considers more and more the cities and the metropolitan areas as continuous territories where agriculture and food are to be included in the urban development. That is how should be understood the evolution between the first international conference "Agriculture in an urbanizing society" held in Wageningen in March 2012, focusing on "Multifunctional Agriculture and Urban-Rural Relations", and the second edition of this conference, held in Roma in September 2015, and dedicated to "Reconnecting agriculture and food chains with societal needs". In this way, scholars have developed systemic approaches of urban-agricultural interactions. They showed for example how these interactions could be strong and provide positive feedback loops for agriculture as well as for the city, as in the African context (Robineau and Soulard 2017). These positive interactions can emerge from urban farmers and inhabitants' everyday practices or proceed from new convergences between urban demands and public supports (Perrin et al. 2013; See chapters, Part III of the book). On the contrary, agricultural-urban dynamics can be mutually exclusive (See Chery et al., Chap. 4).

These perspectives need to adopt a systemic analysis framework linking agriculture, food and urban development, as Wiskerke (2015) proposes to do: by taking a city region perspective, connecting flows, creating synergies, and planning

[4] RUAF: http://www.ruaf.org/

[5] http://www.milanurbanfoodpolicypact.org/

[6] http://sustainablefoodcities.org/

[7] http://terresenvilles.org/

for sustainable urban food systems. Some reflections go further, considering urban regions as ecosystems where to achieve the objective of reconnection and balance between human and physical systems. According to the "bioregionalist perspective" (Magnaghi 2014), like many previous works on urban utopias (Steel 2008), territory is a common: Environment, agriculture, food and rurality are entirely integrated in a renewed social contract (pact) between cities and countryside in the long term. These idealistic visions should not neglect the key role of the actors and power issues in the sustainability of urban food systems.

1.1.3 Positioning Sustainability from the Actors' Perspective

Sustainability is a topic of scholarly debates in urban agriculture and food planning. For some researchers, urban and peri-urban agriculture and local food are social movements enhancing sustainability (Alkon and Agyeman 2011; James 2016): reducing environmental and social inequalities, improving sustainability and democracy, etc. For others, alternative food systems (CSA) and urban agriculture represent new forms of neoliberalism which can reproduce unequal social structures. These scholars consider that contradictions are inherent to urban agriculture (McClintock 2014), which must therefore be analyzed in light of its contribution (or not) to the reduction of inequalities in cities and to the achievement of social justice (Reynolds and Cohen 2016).

Scholarly debates on sustainability of urban agriculture are also to be found in the field of farming systems studies. Scholars distinguish two levels of sustainability (Ba and Aubry 2011; Terrier et al. 2013): (i) internal, or limited sustainability, which is the one of the farm that must be socially livable, economically viable and environmental friendly; (ii) external, or extensive sustainability, related to the contribution of agriculture to the sustainability of the socio-ecological system through urban functions, or ecosystem services. Articulating these two levels of sustainability in urban agriculture is difficult. The analysis should thus also consider the actors' system and public policies to identify trade-offs and to translate sustainability into socially constructed development goals.

This epistemological obstacle led us to propose the concept of "active sustainability" (Soulard et al. 2016). Indeed thanks to standardized and validated indicators, one can quite estimate the sustainability of agricultural systems, according to their performance or their urban functions. However this is another story to try and assess the urban agriculture's sustainability, whether we consider the sustainability of an urban farm (Soulard and Aubry 2011) or the city as a whole (Theys and Emelianoff 2001). It is therefore important to analyze how actors elaborate sometimes contradictory objectives, and how they negotiate trade-offs in the search for sustainability goals. This junction between the diagnosis of a situation and the action-making process can be considered as a key-moment for understanding the sustainable development process "in progress".

This approach of the sustainability based on the consideration of the actors' various points of view gives structure to the book. Part I (Chaps. 2, 3, 4 and 5) highlights, with systemic approaches of the agricultural-urban relations, the fundamental actors-side of the equation of urban agriculture sustainability. In the Part II (Chaps. 6, 7 and 8), researchers design solutions to enhance the sustainability of urban food systems: methods and tools are designed for helping decision makers. In the Part III, Chaps. 9, 10, 11, 12 and 13 deal with the recent transformations in urban agriculture, linking the diversity of farms (urban collective gardens, Zero-Acreage farms, commercial farms…) with the multi-stakeholders governance which is emerging in the cities. All these research works consider the key role of actors in the definition of pathways, which enhance sustainability, seen as an ongoing process.

1.2 The Book: A Sustainability Approach Based on Changing Processes

1.2.1 Systemic Approaches of Agri-Urban Interactions

Part I of the book emphasizes the importance of systemic approaches as a means of highlighting the diversity of relations between agriculture and the city.

Food can be seen as a lever for the reconnection between the city and agriculture. As a matter of fact, many initiatives and policies are based on the objective of relocalization of food. But is agriculture being relocalized? Under what conditions? Sylvie Lardon et al. (Chap. 2) propose an analytical framework (spaces, actors, activities) to describe the urban-agricultural relations around the city of Pisa (270,000 inhab.), Italy. Their results show the range of factors related to the food issue that contribute to link the city and agriculture. Focusing on short food supply chains, Gonçalves A. & Zeroual P. (Chap. 3) provide an economical analysis of food supply, in which the uncertainty of the actors must confront with the level of competition. The result helps to understand why logistics can be organized differently from one chain to another. These elements help also to assess the sustainability of the chains and their adverse impacts. Therefore, developing sustainable food systems for cities through short food supply chains is not only a technical challenge with one-size-fits-all solutions to reduce transport and its impacts, but must consider their complexity and distinctiveness.

The city-agriculture relations can also be understood as a system. J.P. Chéry et al. (Chap. 4) test the agri-urban system as an operating concept for analyzing these relations around the city of Meknès (650,000 inhab.), Morocco. The approach proceeds in three steps and scales: (1) the national system of urban and agricultural policies in Morocco; (2) the local system of relations between the city and the farmers; (3) two drivers of the system change: liberalization of the land agricultural market and housing policy. Unlike the previous chapters, this work shows that the integration of city-agriculture relations comes mainly from non-voluntary mechanisms. These factors can over-determine local strategies of actors. In another context, A. Daburon (Chap. 4) shows that the sustainability of urban dairy producers in the city of Cairo (7,772,000, inhab.), Egypt, depends on several primary constraints: limited space and access to fodder, increases in feed concentrate prices, and manure management. Despite a difficult context, producers maintain their activity by serving the community and taking advantage of its proximity: they incorporate adaptive strategies for feeding systems, breeding systems, and inputs sourcing, and providing employment, food security, and waste recycling. This is a matter of permanent adjustments, with a limit beyond which exiting from agriculture is the only available option.

1.2.2 Methods and Tools to Design Sustainable Urban Food Systems

Researchers are expected to design solutions to enhance the sustainability of urban food systems. This is the core of the reflection drawn in the Part II of the book. Three Chaps. (6, 7 and 8) elaborate and test maps and tools that could be useful for actors engaged in local projects and policies.

Taking the example of the Rotterdam Metropolitan Region, Washer D. et al. (Chap. 6) define the relevant boundaries of a sustainable foodshed. Seeking to optimize the "food miles" of urban supply, they put forward a spatial-analytical approach that combines an ecological footprint with the social and environmental dimensions of the agro-food sector. Their background is based on the adaptation of the Von Thünen model to a dynamic interpretation of the regional food supply potential. They calculate the number of agricultural hectares necessary to feed the city, to be compared with the current land use in the region. It is a tool for communication, raising awareness about the environmental impact of our food supply, as well as a tool to assist in the decision-making of land use allocations in planning. From the point of view of urban planning, Callau S. et al. (Chap. 7) discuss local food systems as a strategic policy at the metropolitan level, not only because of their environmental or food security assets, but also for their contribution to more resilient rural-urban systems. Strategies adopted in Barcelona (Catalonia, Spain) to apply the precepts of ecological urbanism in what is called the "superblocks", are considered and used as a basis on which ensure the development of food planning, the design of urban and food cells, and the optimization of the flow of food between the two. By doing this, this paper highlights the unsustainability of current transportation and food distribution systems in the city, and raises major challenges in co-distribution and concentration of flows.

Based on project-scale approach and relying on the concept of "integrating project", Salma Loudiyi et al. (Chap. 8) describe the various forms of integration of cities and agriculture through a collection of projects case studies found in five cities located in the Mediterranean. Following a theoretical framework emphasizing the role of actors and types of governance, it categorizes the various spatiality and multi-dimensionality of these projects. This perspective of a diversity of projects integrating agricultural and urban dynamics reveals the multiple social, spatial and institutional forms that they are likely to take. The typology of these projects shows an interesting panorama of the various forms of answers to agri-urban sustainability issues by the local projects.

1.2.3 Inventory of Changes in Urban Agriculture

Part III analyzes current changes in urban agriculture. Indeed, in the recent years, the interest for urban agriculture has grown rapidly in Europe. This continuous dynamic intensifies, bringing to light new forms of agriculture, led by new actors, and involving new public supports and expecting to a more inclusive governance. For scholars, that means to compare local situations and case studies. It also drives researchers in action-research experiences with local actors. The five chapters of this section are case-studies of these recent transformations in urban agriculture.

C. Aubry & A.C. Daniel (Chap. 9) differentiate three forms of urban agriculture in the Paris city- region: on-the-ground farms, "low-tech off-soil" systems with open-air cropping systems on rooftops, and "high-tech" greenhouse systems. They show that project leaders have widely diverse background; they give a precise assessment of their contribution to the city's food supply, which is still quantitatively small. They also examine the challenges these farm projects have to face, as difficulties in securing access to open spaces or buildings, or the current suspicion about the safety of agricultural products grown inside a reputedly polluted urban region. In the same way of assessment, S. Thomaier (Chap. 10) provides an inventory of urban farming projects, classified under the umbrella term "Zero-Acreage Farming" (ZFarming): open-air rooftop farms, rooftop greenhouses, productive facades, and indoor farming on and in existing or newly built urban structures. Taking examples of projects found in New-York and Berlin, she analyzes the opportunities and challenges for this type of urban agriculture: site-related requirements, legislative frameworks as well as specific economic risks and opportunities. She also gives an overview of actors involved in planning and implementation processes, stressing the role of landowners and developers, and deducing implications for policies and stakeholder management.

The two following chapters deal with the rise of urban gardening. T. Anthopoulou et al. (Chap. 10) explain the relation between the economic crisis in Greece and the recent growing number of municipal allotment gardens in Athens. According to them, both the local authorities and citizens embrace municipal allotment gardens as alternative spaces within city neighborhoods for ensuring livelihoods and providing a way out of the multiple effects of the recent crisis. Either driven by the economic crisis, or by other motivations such as community building, psychotherapy, re-connection with nature and greening the city, urban gardening projects play an important role in the city as well as in the creation of new identities and a sense of belonging for urban dwellers. P. Scheromm & G. Mousselin (Chap. 12) compare the proliferation of urban collective gardens in Montpellier (France) and Lisbon (Portugal). This comparison addresses the specificity of these gardens, their dynamics and their governance methods. The proliferation in these two cities reflects a strong demand by urban residents, and the support of the municipalities.

The last chapter highlights how multifunctional land use is gaining the attention of urban actors in the city of Zurich, including urban gardeners, farmers and policy-makers as well as city administrators. I. Jahrl & O. Schmid (Chap. 13) show that there are different interests for land use between, but also within, the various urban actors. In particular, farmers partly consider the diverse land use functions as conflicting objectives within the multifunctional concept. Actors with similar land use interests may compete for control of land due to their different forms of organization such as professional urban gardening initiatives and farmers, as well as more hobby oriented gardening initiatives and allotment gardeners. This new and potentially conflicting situation illustrates the need for an integrated urban approach of land use planning and territorial governance towards multifunctional land use in cities.

***

Research works presented in the book are recent. For most of them, research programs date from less than 10 years ago. Therefore this book provides a perspective of ongoing processes in the role of agriculture in the transition towards sustainable urban food systems. It can serve as a milestone for future works, which will likely show new trends and, hopefully, stronger sustainable solutions.

Acknowledgements The authors acknowledge the co-editors of the "Urban Agriculture" series and the reviewers who help to improve the scientific quality of the book. The results of some chapters originate from the DAUME research project funded by the French National Research Agency through the n° ANR-2010-STRA-007-01. Some others chapters originate from presentations at the 5th annual Conference of the Sustainable Food Planning group of AESOP, heled in Montpellier (France) in October 2013.

References

Alkon AH, Agyeman J (2011) Cultivating food justice: race, class, and sustainability. MIT Press, Cambridge

Aubry C, Ramamonjisoa J, Dabat MH, Rakotoarisoa J, Rakotondraibe J, Rabeharisoa L (2012) Urban agriculture and land use in cities: an approach with the multi-functionality and sustainability concepts in the case of Antananarivo (Madagascar). Land Use Policy 29(2):429–439

Ba A, Aubry C (2011) Diversité et durabilité de l'agriculture urbaine: une nécessaire adaptation des concepts? Norois 4:11–24

Bryant C, Charvet JP (2003) La zone périurbaine: structure et dynamiques d'une composante stratégique des régions métropolitaines. Revue Canadienne des Sciences Régionales 26:241–250

Chaboud G, Bricas N, Daviron B (2013) Sustainable urban food systems: state of the art and future directions. In: Presentation, 5th Aesop sustainable food planning conference, Montpellier, 28–29 October

Cohen N, Ilieva RT (2015) Transitioning the food system: a strategic practice management approach for cities. Environ Innov Soc Trans 17:199–217

Cummins S, Macintyre S (2002) A systematic study of an urban foodscape: the price and availability of food in greater Glasgow. Urban Stud 39(11):2115–2130

Duchemin E, Wegmuller F, Legault AM (2010) Agriculture urbaine: un outil multidimensionnel pour le développement des quartiers. VertigO-la revue électronique en sciences de l'environnement 10(2). URL: http://vertigo.revues.org/10436; https://doi.org/10.4000/vertigo.10436

Gottlieb R, Joshi A (2010) Food justice. MIT Press, Cambridge

Hamilton AJ, Burry K, Mok HF, Fiona Barker S, Grove JR, Williamson V (2013) Give peas of chance. Urban agriculture in developing countries: a review. Agron Sustain Dev 34:45–73. https://doi.org/10.1007/s13593-013-0155-8

Heynen N, Kurtz HE, Trauger A (2012) Food justice, hunger and the city. Geogr Compass 6(5):304–311. https://doi.org/10.1111/j.1749-8198.2012.00486.x

Horst M, Gaolach B (2015) The potential of local food systems in North America: a review of foodshed analyses. Renew Agric Food Syst 30(05):399–407

James S (2016) Farming on the fringe. Peri-urban agriculture, cultural diversity and sustainability in Sydney. Springer, Cham

Magnaghi A (2014) La biorégion urbaine: petit traité sur le territoire bien commun. Eterotopia, Paris

Mansfield B, Mendes W (2013) Municipal food strategies and integrated approaches to urban agriculture: exploring three cases from the global north. Int Plan Stud 18(1):37–60. https://doi.org/10.1080/13563475.2013.750942

McClintock N (2010) Why farm the city? Theorizing urban agriculture through a lens of metabolic rift. Camb J Reg Econ Soc 3(2):191–207

McClintock N (2014) Radical, reformist, and garden-variety neoliberal: coming to terms with urban agriculture's contradictions. Local Environ 19(2):147–171

Moragues-Faus A, Morgan K (2015) Reframing the foodscape: the emergent world of urban food policy. Environ Plan A 47(7):1558–1573. https://doi.org/10.1177/0308518X15595754

Morgan K (2009) Local and green, global and fair: the ethical foodscape and the politics of care. Environ Plan A 42(8):1852–1867. https://doi.org/10.1068/a42364

Morgan K, Sonnino R (2010) The urban foodscape: world cities and the new food equation. Camb J Reg Econ Soc 3(2):209–224

Moustier P (2017) Short urban food chains in developing countries: signs of the past or of the future? Nat Sci Soc 25(1). https://doi.org/10.1051/nss/2017018

Perrin C, Jarrige F, Soulard CT (2013) L'espace et le temps des liens ville-agriculture : une présentation systémique du cas de Montpellier et sa région. Cah Agric 22(6):552–558. https://doi.org/10.1684/agr.2013.0671

Pothukuchi K, Kaufman JL (1999) Placing the food system on the urban agenda: the role of municipal institutions in food systems planning. Agric Hum Values 16(2):213–224

Reynolds K, Cohen N (2016) Beyond the Kale: urban agriculture and social justice activism in New York City, vol 28. University of Georgia Press, Athens

Robineau O, Soulard C (2017) Comprendre la complexité des liens ville-agriculture : intérêt d'une approche par le système agri-urbain. Le cas de Bobo-Dioulasso, Afrique de l'Ouest. Nat Sci Soc 25(1) Online. https://doi.org/10.1051/nss/2017013

Sonnino R (2014) The new geography of food security: exploring the potential of urban food strategies. Geogr J 182(2):190–200. https://doi.org/10.1111/geoj.12129

Soulard CT, Aubry C (2011) Cultiver les milieux habités: quelle agronomie en zone urbaine. Revue Agronomie Environnement Sociétés 2(8):89–101

Soulard CT, Valette E, Perrin C, Abrantes P, Banzo M, Bendjaballah O et al (2016) Durabilité des agricultures urbaines en Méditerranée. Rapport final Projet DAUME (ANR-10-STRA-007), INRA, 180 p

Soulard CT, Valette E, Perrin C, Abrantes PC, Anthopoulou T, Benjaballah O et al (2017) Peri-urban agro-ecosystems in the Mediterranean: diversity, dynamics, and drivers. Reg Environ Chang:1–12. https://doi.org/10.1007/s10113-017-1102-z

Steel C (2008) Hungry city. Chattoo and Windus, London

Terrier M, Gasselin P, Le Blanc J (2013) Assessing the sustainability of activity systems to support households' farming projects. In: Marta-Costa AA, Soares da Silva ELDG (eds) Methods and procedures for building sustainable farming systems. Springer, Dordrecht, pp 47–61

Theys J, Emelianoff C (2001) Les contradictions de la ville durable. Le débat 113(1):122–135

Viljoen A, Bohn K (2014) Second nature urban agriculture: designing productive cities. Routledge, New York

Viljoen A, Wiskerke JS (eds) (2012) Sustainable food planning: evolving theory and practice. Wageningen Academic Publishers, Wageningen

Viljoen A, Bohn K, Howe J (2005) Continuous productive urban landscapes: designing urban agriculture for sustainable cities. Elsevier/Architectural Press, Amsterdam/San Diego

Walker RE, Keane CR, Burke JG (2010) Disparities and access to healthy food in the United States: a review of food deserts literature. Health Place 16(5):876–884. https://doi.org/10.1016/j.healthplace.2010.04.013

Wiskerke JS (2015) Urban food systems. In: de Zeeuw H, Drechsel P, HRH The Prince of Wales (eds) Cities and agriculture. Developing resilient urban food systems. Routledge, London, pp 1–25

Part I
Systemic Approaches of Urban-Agricultural Relations

Chapter 2
Food, Integrating Urban and Agricultural Dynamics in Pisa, Italy

Sylvie Lardon, Marie Houdart, Salma Loudiyi, Rosalia Filippini, and Elisa Marraccini

Abstract Food concerns are reshaping the links between urban and agricultural dynamics. They are emerging as significant issues at the interface between public policies and local initiatives that are in turn establishing new approaches in urban planning and land use. Food occupies a concomitant position as an agricultural product of periurban areas, the principal commodity of a supply system, and a focus of community projects. But does this mean that it acts as an integrating element in an agricultural system? Drawing on data from the research project, ANR-DAUME, which focuses on the sustainability of urban agriculture in Mediterranean cities, we analyse the urban agricultural system of Pisa, Italy, in terms of a geo-agronomical model based on a triad of actors, activities, and spaces. This model highlights the diversity of agricultural food production, the hybridization of sales, distribution networks, and the multifunctionality of organizations involved in various related food projects. These aspects combine in a web of relationships that infuse Pisa's urban agricultural system with new abilities to transform and adapt to evolving dynamics, showing that food can play the role of a common denominator, integrating actors from these diverse worlds and transforming the dynamics that influence land use and development in the territory.

S. Lardon (✉) · M. Houdart · S. Loudiyi
University of Clermont Auvergne, AgroParisTech, Inra, Irstea, VetAgro Sup, UMR 1273 Territoires, Clermont–Ferrand, France
e-mail: sylvie.lardon@agroparistech.fr; marie.houdart@irstea.fr; salma.loudiyi@vetagro-sup.fr

R. Filippini
University of Clermont Auvergne, AgroParisTech, Inra, Irstea, VetAgro Sup, Territoires, Clermont–Ferrand, France

Institute of Life sciences, Scuola Superiore Sant'Anna, Pisa, Italy
e-mail: rosalia.filippini@gmail.com

E. Marraccini
Institute of Life sciences, Scuola Superiore Sant'Anna, Pisa, Italy

UniLaSalle, UP2012-10-103 INTERACT, Beauvais, France
e-mail: elisa.marraccini@unilasalle.fr

C.-T. Soulard et al. (eds.), *Toward Sustainable Relations Between Agriculture and the City*, Urban Agriculture, https://doi.org/10.1007/978-3-319-71037-2_2

Keywords Food · Actors · Activities · Spaces · Project · Territory

2.1 Introduction: Food as a Focus of Concern

In the space of a decade, food has gained visibility in developed countries in both the scientific community and the public debate. Numerous scientific works now consider it to be an issue for a combination of public policies. This suggests that food concerns would be appropriate subjects in a transversal treatment of these policy issues and particularly relevant to the question of a transition towards sustainable systems. Morgan (2009) asserts that in the present context, food represents an important element in a variety of sectors including public health, social justice, energy, water, transportation and economic development. Clearly, food is both an individual and collective question that resides at the interface of several sectors. The debate in this domain is currently focused on the question of food security (Sonnino 2014), which leaves the agricultural sector a separate task of reconnecting and addressing the questions of health, social justice, environment, and waste recycling, just to name a few. Today, this transversality is being constructed through the intervention of a diversity of actors exploring food geography, where a stronger relationship between actors will be imperative, whether they are based in the spheres of local government, the market or civil society (Wiskerke 2010; Renting and Wiskerke 2010; Lamine et al. 2012).

Food issues stand at the interface between new public policies and local initiatives that establish new forms of organisation. These issues are reconfiguring the links between urban and agricultural dynamics (Lardon and Loudiyi 2014). This leads us ask how food issues contribute to the integration of urban agricultural systems, defined by Valette et al. (2012) as "the ensemble of relationships and interactions between actors, activities, and urban and agricultural areas, that contribute to the inclusion of agricultural exploitations and other forms of agriculture in a sustainable form of urban development".

In this chapter, we analyse the potential for integration of the Pisa's urban agricultural system by food. This Italian case study is interesting because of the diversity of existing farming systems and the increasing public interest in local food systems. We will describe our analytical framework based on the triad of actors, activities, and spaces and then present the methodological path developed to identify their ways of integration. If we acknowledge food as a product of periurban agricultural production, supply chain, and collective projects, can it then be considered as an integrating element in Pisa's urban agricultural system?

2.2 Analysis Through a Triad of Actors, Activities, and Spaces

The triad of actors, activities, and spaces represents the different dimensions of integration that are systematized to respond to territorial concerns and issues (Benoît et al. 2006). This geo-agronomic model (Lardon 2012) facilitates the understanding and mastering of territorial dynamics from the perspective of local development (Deffontaines et al. 2001). It has already been used to observe, understand, and act on the agricultural dynamics in Tuscany (Rizzo et al. 2012). We use the model here to examine the position of food in the urban agricultural system of Pisa.

Urban agricultural systems are territorial forms of organization that coordinate the interactions of actors, activities, and spaces. Soulard (2014) proposes three versions:

- The urban agricultural system is viewed as part of an agricultural system residing in the urban region. Here we analyse it through the differentiation of the spaces, both in terms of activities they support, and perceptions of the actors, which reveal a symbolic spatial appropriation (Ménadier and Michelin 2012). Land use determines crop management. Field sizes determine allocations of land. Zoning regulations limit the type of farming, but multi-usage practices reinforce a spatial and temporal overlap of activities. The symbolic appropriation reveals trendy or unappealing areas that can change suddenly depending on the location (degraded areas at the periphery of cities, interest in natural areas on the outer limits etc.).
- The urban agricultural system is viewed as a part of the urban system and its activities. These activities are numerous and relate to the dynamics involving residents and tourism, economic and cultural development, and social and spatial segregation. The spatial organization models developed by Lardon et al. (2010) point to the different functional relationships between agricultural and urban activities, depending on their proximity to urban centres. This agriculture is multifunctional and responds to production issues, as well as environmental issues such as water quality and the variety of landscapes, or on a wider territorial level, the conservation or creation of green space (Marraccini et al. 2012).
- The urban agricultural system is viewed as a function of actor involvement. Here we examine the diversity of actors associated at different scales of space and time. These are both individual and collective actors that intervene through different activities, such as the farmers or other professions implicated in the land management (forestry, hydrology, etc.). There are also institutional actors that use zoning mechanisms and land use planning to control the development of various activities in the areas (agricultural, residential, economic, etc.). They work at various levels of organizations leading to interventions that transform the landscape and the urban region (Planchat 2010).

Food contributes to these different descriptions of the urban agricultural system. We put forward the hypothesis that food is indeed an integrating element among the different activities where food acts as a common denominator in the areas of production and consumption, both rural and urban, and among the actors in the sense that they are linked through a common element.

2.3 Case Study and Methodological Itinerary

Pisa is located on the Mediterranean coast of Tuscany, Italy. Seven municipalities constitute Pisa's urban region: Cascina, Calci, Pisa, San Guiliano Terme, Vicopisano, and Buti. With the exception of Buti, the six municipalities form the metropolitan area known as "Area Pisana". This area has a dense population (400 inhabitants/km^2) which totals approximately 200,000 inhabitants, half of whom reside in Pisa. The urban region extends across 500 km^2, and is geographically delineated by the coast on its eastern border, and Monte Pisano to the north. The plain of Valdarno is on the eastern flank which opens onto the region of Florence, and to the south are the hills of the Leghorn province. The coastal plain of Pisa's urban region is planted in cereal and industrial crops, along with livestock and vegetables, and the adjacent hills (900 m.a.s.l.) are almost exclusively devoted to olive oil production. The plain and hills are connected by a complex hydrological system composed of two subsystems: a reclaimed area and a bench terrace-based drainage area (Fig. 2.1).

Thus the urban region of Pisa constitutes a territory that is both an administrative entity, and a geographical entity, defined by a plain and mountain that present a diversity of agriculture. In an analysis of actor perceptions, plans, and documents from regional and local policies, Marraccini et al. (2013) show that there are several issues at stake in Pisa's periurban agriculture. The most frequently cited issues centre around water quality and access, food production, and the protection of farmland from urbanization. Pisa's urban region reflects the general dynamics of urbanization in agricultural areas close to urban centres, and competitive challenges for small scale farming activities. The average farm size in the area was 8 ha according to the last agricultural census (ISTAT 2010).

The methodological itinerary (Fig. 2.2) developed in the framework of the ANR DAUME[1] project was undertaken between 2011 and 2013. During a preliminary phase conducted in 2011, we identified urban and agricultural issues through an analysis of urban and rural planning documents, and an interview process with approximately two dozen local actors (Marraccini et al. 2013).

Different sources of information were combined to identify the activities (document analysis, interviews with diverse actors), characterize the spaces (analysis of satellite images) and involve the actors (participatory workshops). One group of documents covered institutional records on urban policies. Other documents concerned projects either identified by actors or through a search of internet sites. Interviews were conducted with different actors. An in-depth questionnaire on farming systems was given to 55 professional farms. They were identified from existing databases (Land Parcel Identification System or LPIS, agricultural organi-

[1] http://www1.montpellier.inra.fr/daume/

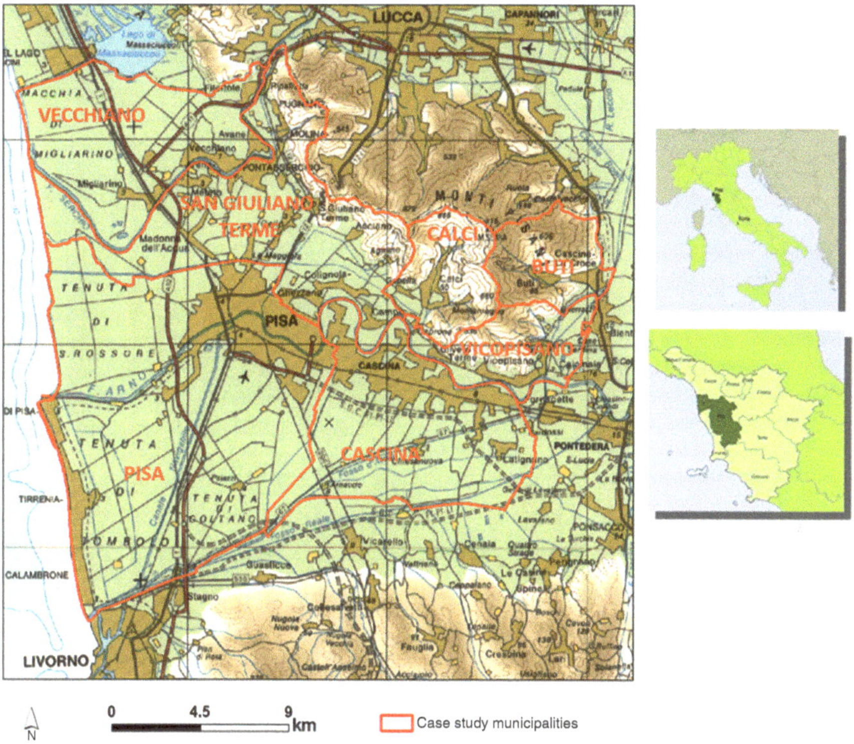

Fig. 2.1 Urban region of Pisa (Tuscany, Italy) (Source: Institute of Life Sciences; Tuscan region geographical database)

zations, Pisa Chamber of Commerce), and sampled on the basis of their main crops, their size, and their spatial distribution (they represent 15% of total farms in the LPIS database in 2012). We also conducted semi-structured interviews among a sampling of hobby farmers in one of the municipalities (Calci, 35 individuals, representing 12% of the hobby farmers identified in the census), to better understand their contribution to periurban agriculture. In addition, 15 interviews with commercial actors provide information on marketing methods which complements the interviews with farmers. Representatives from the region, the province, and municipalities, as well as local authorities (basin syndicate, land reclamation authority, etc.), and intermediate actors (agronomists, technicians, associations, etc.) were questioned about specific issues relating to local governance methods and the regionalization of public policies. An analysis of Landsat and Spot images, combined with a partial SIG database enabled a reconstruction of the evolutionary dynamics of land use in the urban region of Pisa (Marraccini et al. 2015). The various data produced were reviewed with the actors and validated over the course of a diverse group of participatory workshops held in 2014 and 2015 (Lardon et al. 2016). A total of more than 150 actors were involved in the questionnaires, interviews, or participatory workshops.

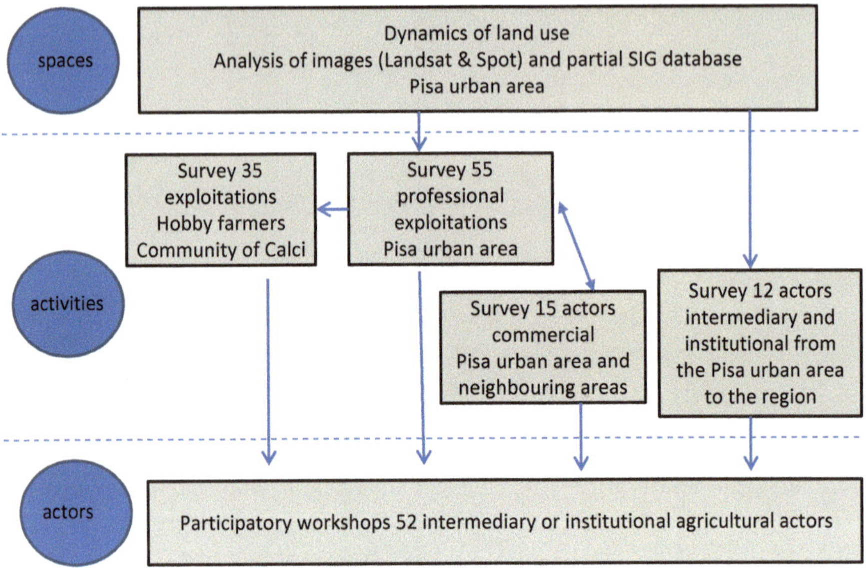

Fig. 2.2 Methodological itinerary of the ANR DAUME project

With respect to our research question concerning food, the information collected was analysed through the lens of the triad of actors, activities, and spaces. The spatial analysis of agricultural land use was developed both on the scale of Pisa's urban region and on the scale of professional farms. The analysis of activities centres on the production and marketing of local food products. The analysis of the actors examines the collective projects uniting urban and agricultural actors around the subject of food.

2.4 Food in the Urban Agricultural System of Pisa

The element of food spans the various components of Pisa's urban agricultural system. It is dependent on the ways in which agriculture invests in periurban areas and responds to environmental and territorial issues. Food is the product in the chain that links the supply of agricultural production with urban demand. It is also the material of initiatives and collective projects. But does that mean that food is an integrating element among the spaces, activities, and actors of Pisa's urban agricultural system?

2.4.1 *Food Production in a Patchwork of Urban and Agricultural Areas*

In Pisa's urban agricultural system, the tangled patchwork of agricultural fields and urban or natural areas can be observed from two organizational perspectives, that of an urban area or that of a farm. This mix of urban and agricultural land uses is the result of 30 years of urban dynamics that created a fragmented image of farm fields where a variety of farming systems have been developed.

2.4.1.1 Dynamics of Land Use: A Decrease in Agricultural Surfaces with Benefits Going to Urbanization and the Expansion of Natural Wooded Zones

An analysis of the evolution in land use between 1985 and 2011 (Marraccini et al. 2015) shows that the urban area around Pisa was subjected to a twofold process reducing agricultural land use (Fig. 2.3). Urbanization spread eastward along the axis between Pisa and Florence, and around the urban centres.[2] This urbanization occurred at the expense of agricultural surfaces. At the same time, natural woodland expanded in some of the areas where agricultural activities were abandoned, particularly in the northeast area of Monte Pisano. In this area where 21% of the agricultural land reverted to wilderness, olive orchards were the most visible victim.

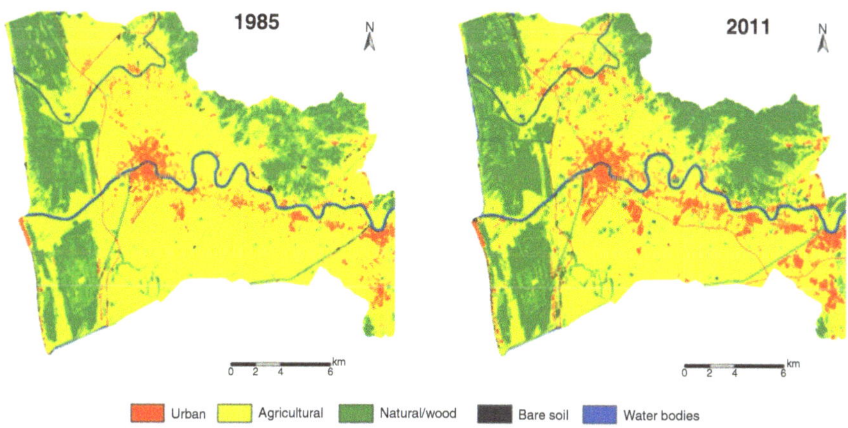

Fig. 2.3 Dynamics of land use in the urban area of Pisa between 1985 and 2011 (Source: Institute of Life Sciences, elaboration of Landsat data)

[2] Even though urban planning policies have somewhat contained the dispersion of actual construction around the city of Pisa itself, contrary to the rest of urban coastal area, because of the presence of a regional natural park.

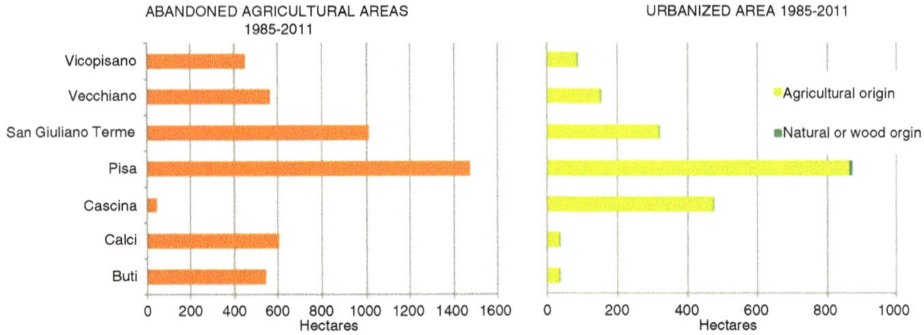

Fig. 2.4 Surface of abandoned agricultural areas and urbanized areas developed on land that was previously agricultural or natural, arranged by the different municipalities of Pisa's urban area (Source: Institute of Life Sciences, elaboration of Landsat data)

Thus, not only did agricultural areas contract on the plains subjected to development pressure from urbanization, but also in the region of Monte Pisano where natural wooded areas spread across abandoned agricultural land (Fig. 2.4).

Could unused interstitial land be reincorporated into an agricultural system in sectors where dense infrastructure still leaves the surface accessible? Indeed, the protection of agricultural land is integrated in numerous institutional planning documents (Marraccini et al. 2013) and was the subject of a regional law establishing a land bank to manage abandoned or under-exploited agricultural terrains. Could it be used for local food production?

2.4.1.2 Distribution of Agricultural Exploitations: Fragmentation and Distance from the City

Our survey of professional farms in the urban area reveals the principal orientation of local farming systems[3]: 65% cereal and industrial crops, 14% livestock, 13% vegetable crops, and 8% olive groves. The farm size varies depending on the farming system, ranging from a few hectares for olive production to 250 ha for farms of industrial crops such as corn or sunflower. High values of the indicators of fragmentation (Fig. 2.5) reveal the significant spatial dispersion among the fields of the various farms (up to an average of six blocks of parcels for farms of industrial crops). Parcels of the same category of farms are found more or less separated (up to a distance of 14 km in the case of cereal farms), and they are found at varying distances from the centre of Pisa (up to 3.5 km for livestock exploitations). The spatial dispersion of olive groves is significant (4 km between blocks) even though they represent limited presence in terms of farmland. Market farms producing vegetables also appear relatively far (more than 2 km) from the urban centres considering their traditional link to the urban market, and their degree of dispersion (more than four blocks).

[3] Defined by the most important crop in terms of surface area of cultivation.

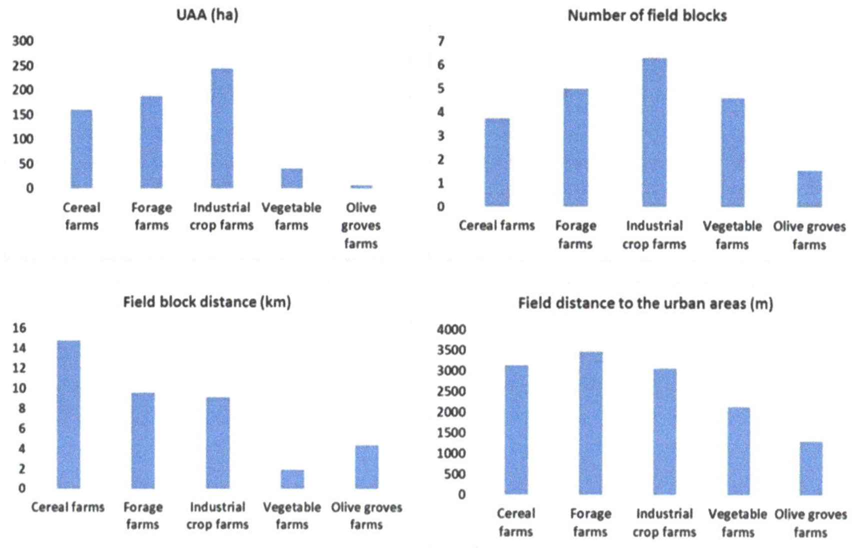

Fig. 2.5 Indicators of fragmentation in the surface area among 55 farms surveyed in 2012/2013 (Source: Institute of Life Sciences elaboration of ARTEA-LPIS data and 2013 survey). UAA stands for Usable Agricultural Area, forage farms were mainly livestock oriented farms

The fragmentation evidenced here is linked to urbanization and infrastructure. It represents a constraint in farm management that requires a reorganization of crop rotation, and a modification of farming practices. This also interferes with the numerous and diverse issues related to water in the territory (Debolini et al. 2007). Water management officials (improvement consortium, basin syndicate) have suggested that changes in farming practices and the transformation of farming systems towards local agricultural forms are possible solutions that could limit flood risks that affect the coastal plain. Could local food production from the development of agriculture in these fragmented areas respond to these environmental and territorial issues? How can food articulate local activities?

2.4.2 *Food, Between Agricultural and Urban Activities*

In fact, it appears to be a strong demand for local products from Pisa's urban agricultural system. The supply of products is diversified and many farmers have embraced marketing strategies that are oriented towards local networks. This match between agricultural supply and urban food demand is also visible in the development of agricultural activities by urban actors, and even in the development of leisure activities developed by farmers and other agricultural actors.

2.4.2.1 Food Supply and Demand

Interviews with local and regional actors indicate that consumers in the urban area of Pisa are passionate about local products. They shop directly with farmers and in stores, where they consistently demand quality local products. Municipalities have established farmers markets; the supermarkets and restaurants have responded to demand with a commitment to sell local products. Both store owners and farmers suggest that the local demand is strong enough that there is little competition. In light of these observations, could the urban region of Pisa represent a consumer market for local products that can be further exploited by farmers?

Across the table from this demand, the different farms in the urban region offer a variety of local products (meat, fruit and vegetables, olive oil, eggs, milk and cheese, bread, and fodder for livestock). Cereals are the only local products that are not directly sold on the local market – except as a transformed product in bread. In addition, the current trends favour diversification and hybridization of food supply networks, which has finally resulted in the dedication of a significant part of production to local distribution (Filippini 2015). So, among the 55 farmers surveyed, 26 are concerned by local market: five exclusively through local networks, and 21 through diverse marketing efforts, 16 of them currently marketing more than 50% of their production through local supply networks.

The existence of cooperatives enables small farmers to participate in larger supply networks; this provides a level of financial security to an otherwise limited farming. The supermarkets diversify the food supply network by carrying lines of local products as a response to consumer demand. The diversity in food supply networks facilitates the development of complementary farmer markets, and also contributes to the sustainability of farmers that are trying to diversify in local products. Nevertheless, the market for local products in the region is far from being saturated. In an analysis of beef and lamb from periurban farm production that was sold in the local food supply chain, Filippini et al. (2014) were able to demonstrate that most of the beef production was labelled local and sold locally, whereas the lamb which was local but not labelled as such, was marketed through conventional food chains. This suggests that the potential supply of locally produced food depends not only on the effective size and number of farms, but also on the collective actions taken regarding local production. Do the different actors in the food supply chain have the ability and the will to modify their actions?

2.4.2.2 The Weave of Activities Surrounding Food

The diversification of activities is not limited to local food production. Many of the farms have multiple activities, with offerings that respond to urban demand for both agricultural food products, and leisure activities. The surveys, from both professional and hobby farms, reveal overlapping activities.

In example, a Monte Pisano farm (Fig. 2.6) combines local food chain and services to valorize its resources. It's an organic olive oil production (transformed

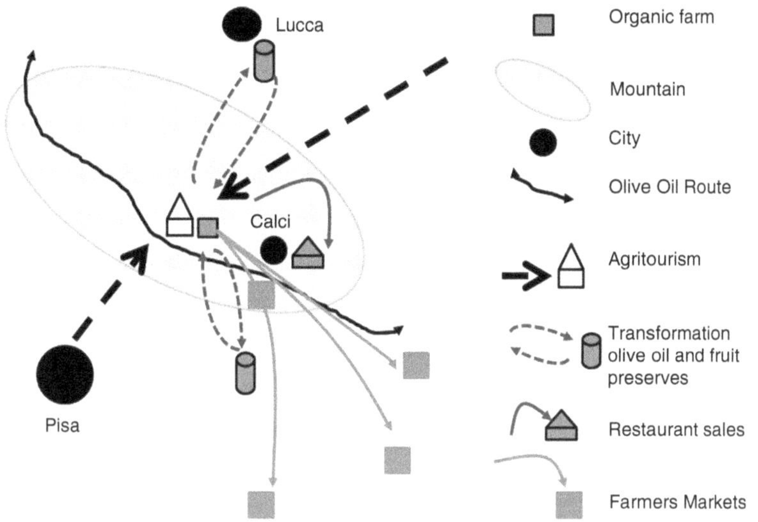

Fig. 2.6 Example of a Monte Pisano farm: diversification of agricultural activities (Source: R. Filippini, elaboration of survey data 2013)

nearby) and fruit preserves (transformed in Lucca) local production farm. The farm markets its products through direct sales on site (70%) further supported by a complementary agritourism activity with a local (Pisa) and international clientele. The balance of production is marketed through a nearby restaurant (15%) or in the farmers markets (15%) organized by the municipalities. This example illustrates how agritourism reinforces the distribution of food products in the urban region of Pisa.

Another innovative cereal farmer-retailer, (Fig. 2.7), combines large conventional supply networks with small local networks. The farmer's sunflower production is sold through the local cooperative which uses conventional supply networks. He distributes 50% of his wheat production through sales to the central distribution in Florence for Tuscan supermarkets. The balance of his production uses the local networks, transforming wheat to flour at a local mill and in his own bakery. This bakery transforms the flour into bread which the farmer then markets through various stores in the city and periurban zones. He established a network of door-to-door sales in his own village, and distributes in another local community where he participates in a project supplying school cafeterias. Finally, he opened a retail store, associated with a bakery in Pisa, where consumers can also find other products from neighbouring farms. This is an example of production, transformation, sales and distribution activities completely interwoven and integrated into various areas within and beyond the urban region of Pisa. It should be noted that the diversity of activities in these local networks is built on the support of the financial capacity provided by conventional supply networks.

Outside of professional farms, some citizens practice agriculture by developing gardens that they cultivate individually or collectively, with or without the support of the municipalities. The importance of these hobby farmers, notably among olive

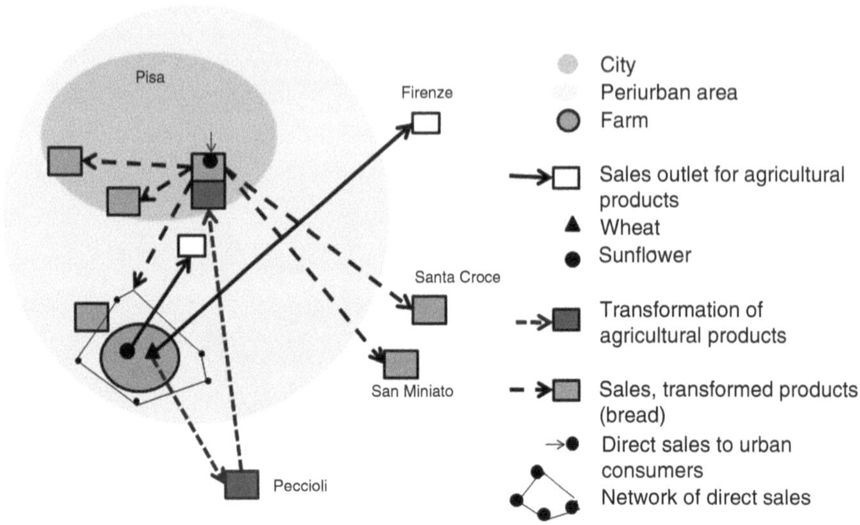

Fig. 2.7 Example of a Coltano (south of Pisa) farm: links between areas of activities (Source: R. Filippini, elaboration of data from surveys 2013)

orchards and the transformation of olive oil in the Monte Pisano area, reflects the strong appeal of agricultural activity among urban actors (Gennai-Schott et al. 2014). Hobby farmers account for 90% of the olive oil producers in the municipalities of Calci. But the owners of these olive groves are often retired (48%), and the olive oil produced is distributed among family and friends. The engagement of these farmers springs from their passion for cultivation and for the landscape, including an awareness that the rehabilitation of the terraced olive groves contributes to the identity of the territory. The protection of these terraces (Rizzo et al. 2007) represents an opportunity for interaction between agricultural and urban actors from the mountain and the plain. These actors recognize the complementary relationship between the mountain and the plain, and emphasize the benefits of this complicity in their vision of the evolution in territorial development.

2.4.3 Food, Prompting New Forms of Coordination between Urban and Agricultural Actors

We can observe that urban and agricultural actors in the region around Pisa have come together and worked together through a weave of activities across a scattered patchwork of areas. New networks of coordination among these actors are emerging in the urban agricultural system, and intermediary actors are facilitating new opportunities for agricultural actors to interact with their counterparts in the urban market.

2.4.3.1 Food, at the Heart of Projects Linking Urban and Agricultural Actors

The identification of projects, directly or indirectly linked to the food system in Pisa's urban region,[4] shows the diversity among participating actors and their relationships.

These projects were established by urban or rural actors, as a public or private initiative. Community Supported Agriculture type producer-consumer associations (GAS in Italy), strongly linked to Pisa Slow Food, and self-production (see CAPS below) were initiated by urban actors, as well as the Pisa's urban gardens even though those were by institutional actors. A number of collective actions link all the actors. These actions support local networks through the mechanisms of territorial projects such as an olive oil route, or the farmers markets organized by some municipalities. These initiatives have been grouped under the *Piano del Cibo* (Food Plan, see below) and promoted by the province of Pisa (Di Lacovo et al. 2013).

These efforts are encouraging new groups of actors from various sectors (urban, rural, public and private) and with various goals (economic, technical, social, cultural, and pedagogical), to form around the issues of food. However, are these dynamics sustainable and reproducible?

2.4.3.2 The Food System Sees the Emergence of Intermediary Actors and Transient Networks

We have noted that the most innovative and successful projects in Pisa seem capable of evolving over time. This suggests that long term sustainability may depend on the project's capacity to adapt or be transformed through the mediation of involved actors. Two examples illustrate this issue.

The CAPS project (Agricultural Community of Social Promotion[5]) corresponds to a collective of citizens in Pisa, Calci, and Vecchiano that was created in 2010. At the time of the interviews, it was an association of several dozen individuals that employed one farmer. They rented 3 ha of farmland from an owner not interested in developing the property, which is on the periphery of Pisa and surrounded by an urban zone. According to the participants, the CAPS project aims to go beyond the traditional relationship between farmers and consumers. They envision resident urban actors participating and applying the principles of social justice and environmental sustainability to agricultural practices and the production of their own food. This model is conceptually positioned somewhere between classical GAS projects and urban gardens, functioning as a hybrid that integrates actors from the urban and rural spheres. However, in 2014, the farmer did not want to continue with the project, because it was too constraining for him to fulfil alone the multiple demands of the consumers. Before seeking a replacement in the original form, the association

[4] See Loudiyi et al. (Chap. 8), in the same publication.

[5] http://www.caps-pisa.org/

decided to experiment with a new formula involving a contract with an existing farmer in the northern part of the territory which is still close to the city, but less interactive with the urban consumers.

Thus the CAPS project no longer exists in the same location, and it did not maintain a relationship with the same farmer, but the collective of actors experimented with an agricultural practice and intend to repeat the experiment at a different location. Access to self-produced food initiated the actors into the agricultural reality of producing food by experiencing the spatial and organizational constraints.

The Piano del Cibo[6] was established with the help of the University of Pisa and elected authorities from the province of Pisa. It was conceived as a tool for building a food strategy on a provincial scale. The plan has been in the making since 2009. It currently acts as a means of making food issues more visible in the urban area of the province by linking civil society with public actors. The plan was officially institutionalized through an act of the province in April of 2010. This was a crucial step in the recognition of a need to build a charter that outlines a common vision and objectives, and details a food strategy with a plan of action. The writing of the charter and the food strategy followed a long participatory process that included a diverse selection of stakeholders: local authorities, the scientific community, health services, economic advisers, associations, and various formal and informal groups. The two documents were presented in October of 2011 with broad support including that of local authorities. However, in spite of the initial enthusiasm, this food plan has not been implemented or integrated in the municipalities as it had been previously anticipated. Nevertheless, it has led to local initiatives, such as one of the first Food Policy Councils in Italy, established at the end of 2013. It fits into an existing institutional slot, the Territorial Council of Participation, a type of neighbourhood council, developing a laboratory of food governance beginning with an exploratory phase mapping the potential for food production in neighbourhoods.

At the time of its conception, *Piano del Cibo* was considered a unifying moment in the discourse on food issues, but there is still little evidence of solid implementation by the municipalities. Nevertheless, it has become a base of support for new urban initiatives and could encourage the future development of local agricultural projects.

These tentative networks may be short-lived or only partly achieved, but they are also instigators of projects in other domains, and they highlight the role of intermediary actors that accompany actions. Beyond the classic actors, such as the natural parks that act in the preservation of agricultural land and the distribution of products, or territorial public actors that support the promotion of local markets and supply networks, other types of intermediate actors are emerging. Universities have a significant role in the education of public and private actors, and serve as intermediaries in the circulation of ideas. Agronomists and civil technicians share both their professional knowledge, and their organizational experience (in regional communities for example). The local and regional authorities are involved in the promotion of the agricultural and tourist industries, the management of risks, and the creation

[6] http://pianodelcibo.ning.com/page/piano-del-cibo-pisa

of new partnerships. All these initiatives could weave links in the territory, organize actors, activities and spaces, and help integrate the various issues.

2.5 Conclusion: Food, an Integrating Element?

In this chapter, our analysis has described the urban agricultural system of Pisa as a system that coordinates actors, activities, and spaces to develop an integrated response to territorial issues. We have questioned whether food acts as an integrating element among these different components.

Our observations show that local food production can counter the dynamics of the abandonment of agricultural land use, and be an integral part of periurban areas in spite of urban development, the intrusion of natural growth, and fragmentation in agricultural land. We have assessed an existing potential of local agricultural production that is sustained through a combination of different activities including food processing, tourism and leisure, and distributed through a hybrid combination of conventional and local supply networks. The research has identified numerous initiatives involving the food system that favour the cooperation of agricultural and urban actors, even if the sustainability of the projects and their impact at regional level are questionable. Our research in the urban region of Pisa suggests that investing in interstitial spaces, combining diverse activities, and mobilizing a variety of actors in an innovative food system offer broad potential that not only responds to the concerns of local food supplies, but can also address territorial issues of the urban agricultural system. Local initiatives and global dynamics involve preservation of agricultural land, water management, and territorial governance for an integrated development.

The dynamics of the territory, the perceptions of the actors, and the concerns of the public, are all converging around the integration potential of food. This being the case, if we envision food as an integrating element, it can act as a development tool for the urban agricultural system of Pisa. Food is not simply playing a direct role as the product in an exchange between agricultural supply and urban demand, it is also an integral part of the responses being applied to a variety of territorial issues. It contributes to the sustainability of the periurban farming systems by offering development alternatives. The diversity of products, the hybridization of food networks, and the multi-functionality of the various organizations create a food web that confers an ability to transform and adapt to the current evolution of the territory.

However, in spite of these positive aspects, food as an element of integration stumbles in two ways. It struggles to become an intermediary element (in the sense of Vinck 2009), at the border between the agricultural and urban worlds. There are certainly spaces, times, and actions that are shared by the actors, but not enough to endure and become a shared goal. It also fails as an integrating element (in the sense of Schmid and Hatchuel 2014) that coordinates the differentiated parts of knowledge and projects itself into the future. There is certainly evidence of an attraction, but it remains out of reach.

Could research on the urban agricultural system of Pisa contribute to the role food is playing as an element of integration between the actors of heterogeneous worlds, to transform the current dynamics into development potential for the territory?

Acknowledgements This analysis is based on research financed by the French National Research Agency through the project DAUME n° ANR-2010-STRA-007-01.

References

Benoît M, Deffontaines J-P, Lardon S (2006) Acteurs et territoires locaux. Vers une géoagronomie de l'aménagement, INRA-Quae, Paris. 176p

Debolini M, Jacopin R, Marraccini E, Pistocchi C, Rizzo D, Toillier A (2007) Territorial diagnosis using spatial representations – which are the relationships between territorial dynamics and agro-environmental issues? The case of the plain of Pisa, Italy. In: Batton-Hubert M, Joliveau T, Lardon S (eds) Proceedings of the international congress on geomatics and spatial analysis. Clermont-Ferrand, France, pp 18–20.

Deffontaines J-P, Marcelpoil E, Moquay P (2001) Le développement territorial: une diversité d'interprétations. In: Lardon S, Maurel P, Piveteau V (dir.) Représentations spatiales et développement territorial. Editions Hermès, Paris, pp 39–56

Di Lacovo F, Brunori G, Innocenti S (2013) Il piano del Cibo. Agriregionieuropa 32:9–12

Filippini R (2015) Food production potential of periurban agriculture: contribution of periurban farms to local food system. Ph. D. Course in Agrobiosciences, Scuola Superiore Sant'Anna (Pisa) & AgroParisTech ABIES (Paris)

Filippini R, Marraccini E, Lardon S, Bonari E (2014) Assessing food production capacity of farms in periurban areas. Ital J Agron 9:63–70

Gennai-Schott S, Rizzo D, Sabbatini T, Marraccini E (2014) Uno studio dell'intensità dei sistemi olivicoli terrazzati: il caso del Monte pisano. XLIII Convegno Nazionale della Società Italiana di Agronomia, Pisa, Italy, 17–19 September 2014

ISTAT (2010) Censimento Generale dell'Agricoltura. Roma

Lamine C, Renting H, Rossi A, Wiskerke JSC, Brunori G (2012) Agri-food systems and territorial development: innovations, new dynamics and changing governance mechanisms. In: Darnhofer I et al (eds) Farming systems research into the 21st century: the new dynamic. Springer, Dordrecht, pp 229–256

Lardon S (ed) (2012) Géoagronomie, paysage et projets de territoire. Sur les traces de Jean-Pierre Deffontaines. Editions QUAE/NSS Dialogues, Versailles, 344 pages

Lardon S, Loudiyi S (2014) Agriculture urbaine et alimentation: entre politiques publiques et initiatives locales. Géocarrefour 89(2):3–10

Lardon S, Galli M, Marraccini E, Bonari E (2010) Peri-urbanisation and peri-urban agriculture: issues and proposals. In: Galli L, Marraccini B (eds) Agricultural management in peri-urban areas. The experience of an international workshop. Felici Editore, Pise, pp 7–24

Lardon S, Marraccini E, Filippini R, Gennai Schott S, Johany F, Rizzo D (2016) Prospective participative pour la zone urbaine de Pise (Italie): l'eau et l'alimentation comme enjeux de développement territorial. Cahiers de Géographie du Québec 20(170):265–286

Marraccini E, Rizzo D, Debolini M, Planchat C, Toillier A (2012) Contribution of agronomy to land management issues. A comparison of five interdisciplinary PhD theses. In: Producing and reproducing farming systems: New modes of organization for sustainable food systems of tomorrow. European IFSA symposium, Aarhus, Denmark

Marraccini E, Lardon S, Loudiyi S, Giacchè G, Bonari E (2013) Durabilité de l'agriculture dans les territoires périurbains méditerranéens – Enjeux et projets agri-urbains dans la région de Pise (Toscane, Italie). Cahiers Agric 22(6):517–525. https://doi.org/10.1684/agr.2013.0658
Marraccini E, Debolini M, Moulery M, Abrantes P, Bouchier A, Chéry J-P, Sanz E, Sabbatini T, Napoleone C (2015) Common features and different trajectories of land cover changes in six western Mediterranean urban regions. Appl Geogr 62:347–356
Ménadier L, Michelin Y (2012) Quand les producteurs d'AOC en Auvergne se penchent sur leurs paysages. Production de formes et de modèles paysagers. In: Lardon S (ed) Géoagronomie, paysage et projets de territoire. Sur les traces de Jean-Pierre Deffontaines. Editions QUAE/NSS Dialogues, Versailles, pp 223–241
Morgan K (2009) Feeding the city: the challenge of urban food planning. Int Plan Stud 14(4):341–348
Planchat C (2010) Agriculture, landscape representations and local urban planning. The case study of the municipality of Billom in France. In: Galli L, Marraccini B (eds) Agricultural management in peri-urban areas. The experience of an international workshop. Felici Editore, Pise, pp 94–104
Renting H, Wiskerke H (2010) New emerging roles for public institutions and civil society in the promotion of sustainable local agro-food systems. In: Transitions towards sustainable agriculture: from farmers to agro-food systems. Presented at the 9th European IFSA symposium, Vienna, 4–7 Juillet
Rizzo D, Galli M, Sabbatini T, Bonari E (2007) Terraced landscapes characterization. Developing a methodology to map and analyze the agricultural management impacts (Monte Pisano, Italy). Int J Geomatics Spat Anal 17:431–447
Rizzo D, Marraccini E, Debolini M, Gallli M, Bonari E (2012) Agronomes et géoagronomie: relecture de trois études de cas en Toscane. In: Lardon S (ed) Géoagronomie, paysage et projets de territoire. Sur les traces de Jean-Pierre Deffontaines. Editions QUAE/NSS Dialogues, Versailles, pp 43–62
Schmid A-F, Hatchuel A (2014) On generic epistemology. Angelaki (USA)
Sonnino R (2014) The new geography of food security: exploring the potential of urban food strategies. Geogr J. Article first published online: 27 Nov 2014. https://doi.org/10.1111/geoj.12129
Soulard C-T (2014) Pratiques, politiques publiques et territoires: construire une géographie agricole des villes. Universite Bordeaux Montaigne, Bordeaux
Valette E, Perrin C, Soulard C-T (2012) Sustainable cities vs sustainable agricultures. A scientific project on agri-urban systems, North and South of the Mediterranean. In: Conference A.U.S., Wageningen, The Netherlands
Vinck D (2009) De l'objet intermédiaire à l'objet-frontière. Revue d'Anthropologie des Connaissances 3(1):51–72
Wiskerke JSC (2010) On places lost and places regained: reflections on the alternative food geography and sustainable regional development. Int Plan Stud 14(4):369–387

Chapter 3
Logistic Issues and Impacts of Short Food Supply Chains: Case Studies in Nord – Pas de Calais, France

Amélie Gonçalves and Thomas Zeroual

Abstract Short food supply chains (SFSCs) are often seen as being capable of making a valuable contribution to the provisioning of cities from an environmentally, economically, and socially sustainable perspective. However, there are doubts as to whether these supply chains deserve this reputation. They involve varied needs and logistical challenges for both farmers and cities, and can have equally varied, and sometimes adverse, economic, environmental and social impacts. These challenges and impacts seem also different from one chain to another. This chapter presents a theoretical framework to help understand these diverse challenges and impacts among SFSCs. Using the economics of conventions, we highlight the importance of coordination between production and demand. This framework is applied to SFSCs operating with one intermediary in the French region of Nord – Pas-de–Calais. Our observations show that the need for rapid and frequent deliveries remains important in SFSCs, making them vulnerable to the negative impacts associated with transport intensive flows of products. The management of information flows varies from one chain to another, and some require more complexity and exchanges between actors. The relationships created by these exchanges can have positive social impacts, creating ties between producers and customers. One type of chain appears particularly suited to the creation of sustainable logistic schemes.

Keywords Logistics · Short food supply chains · Urban areas · Diversity

---

A. Gonçalves (✉)
AGIR, Université de Toulouse, INRA, INPT, INP-EI PURPAN, Auzeville, France
e-mail: amelie.goncalves@inra.fr

T. Zeroual
ESCE, CIRCEE, Paris, France
e-mail: thomas.zeroual@esce.fr

C.-T. Soulard et al. (eds.), *Toward Sustainable Relations Between Agriculture and the City*, Urban Agriculture, https://doi.org/10.1007/978-3-319-71037-2_3

3.1 Introduction

Today there is a renewed interest in the local food supply of cities, and many developed countries have created policies of food governance (Viljoen and Wiskerke 2012) with the aim of developing sustainable, local food systems (Hinrichs 2003). These policies are often based on short supply chains that are assumed to be better from economic, social and environmental perspectives than other types of chains (European Commission 2014).

However, academic literature suggests that being short is not *per se* a guarantee of global sustainability and many questions remain (Aubry and Chiffoleau 2009). Logistics is one of the determinant factors of this phenomenon because the level of efficiency in the circulation of goods and information can have a significant impact on the economic and environmental performance of a supply chain (Jarosz 2008; Coley et al. 2009; Nordmark et al. 2012; Bosona et al. 2013). Being short can also impact social sustainability when logistical issues complicate the work of farmers and undermine their relationship with consumers, whereas this relationship is often represented as one of the main objectives of these supply chains (Herault-Fournier 2013; Lutz and Schachinger 2013).

The literature also shows that these issues are prominent in cities, where high density and difficult circulation increase the logistic challenges, particularly those of transportation for distribution (Morganti and Gonzalez-Feliu 2015; Schliwa et al. 2015). The building of more sustainable food systems to feed cities thus appears to require specific solutions (such as food hubs and integrated logistics) to guarantee the positive economic, environmental and social impacts of short supply chains (Nordmark et al. 2012; Cleveland et al. 2014).

Furthermore, academic papers have also highlighted the wide diversity of logistic schemes in short chains (Blanquart et al. 2010; Aubry and Kebir 2013). This diversity can be seen in the number of actors involved, the geographic scale, the size and type of distribution systems. These works also describe differences in the economic, social and environmental sustainability among chains (Aubry and Chiffoleau 2009; Capt et al. 2011; Hayden and Buck 2012; Mundler and Rumpus 2012).

Why is there such a diversity of logistical organization and performance? Is it possible to develop a theoretical framework that could describe and explain this diversity? Our hypothesis is that this diversity reflects particular types of short food supply chains (SFSCs). Each one has its own needs with respect to the coordination between actors. This coordination explains the logistical organization and its performance.

We will begin by defining short food supply chains and their logistics, and describing the perspective of current academic literature on this logistics and its performance (Sect. 3.1). Following, we will explain our economics of conventions framework, which is based on the question of coordination among actors in the process of supplying products to a targeted demand (Sect. 3.2). Then (Sect. 3.3), we will apply this framework to our study of short supply chains in a French region chosen for its high instance of short chains and significant population density in multiple urban areas, in other words, a region where logistics in short chains may represent a problem: Nord – Pas-de-Calais.

3.2 Logistics: A Primary Issue in the Various Dimensions of Short Food Supply Chains

Because they can be very different from one another, all SFSCs do not have the same logistical needs. In addition, these needs refer not only to the transport but also to other functions (Sect. 3.2.1). Consequently, logistics is one of the key factors of the chain's economic, environmental, and social performance (Sect. 3.2.2).

3.2.1 Logistics in Short Food Supply Chains: A Diversity of Transportation Needs... But Not Only

Logistics is a function whose purpose is to enable companies to meet their objectives by coordinating the flow of goods, information and money throughout the supply chain (Savy 2007). The latter can be defined as the path of goods or services as they travel from production to consumption (Lendrevie et al. 2009: 371). Thus, logistics is a function dedicated to the management of physical flows such as ordering, transport, or storage. However, discussing logistics also means discussing how information is exchanged throughout the chain (the means, the frequency and the type of information).

SFSCs distribute processed or unprocessed products with few or no intermediaries between the producer and the customer (Parker 2005). A geographical criterion is frequently added to this definition (Feagan 2007) although there is no consensus regarding the number of kilometres it allows. Additionally, these chains are often differentiated on the basis of an ethical distinction (Deverre and Lamine 2010; Williams et al. 2015) and a response to the desire for more personal relationships between actors than those found in classical retailing channels.

Therefore, SFSCs refer to different types of chains. They can vary from the shortest and simplest chains with no intermediary, short distances and the potential for strong ties between actors (Community Supported Agriculture, CSA, box schemes, or farmers' markets for example) to the most complex. The latter are multi actor solutions with many customers and at least one intermediary such as a local procurement for supermarkets or restaurants and public catering. These chains can also involve many actors and occur in large geographical areas (Ljungberg et al. 2013). The smallest chains are based on very simple logistics with few flows to manage (Blanquart et al. 2010). The more complex chains need more sophisticated and optimized organization such as platforms (Bosona et al. 2011) and/or the intervention of a service provider (Martikainen et al. 2014) to manage important needs regarding transport, storage, and "administrative tasks" such as order management.

These issues are not the same in rural and urban areas. Access to large cities is particularly difficult for farmers who must contend with long distances and complicated circulation (Morganti and Gonzalez-Feliu 2015). These problems reflect the diversity of short food supply chains in terms of logistics, transportation, customer relationships and order management. However, regardless of the level of complexity or efficiency, these chains and their logistical functions have economic, social and environmental impacts.

3.2.2 Logistics as One of the Key Factors of SFSCs' Performance

Economically, in terms of profit, some short chains face additional costs (Jarosz 2008) for transport (Nordmark et al. 2012) and marketing (a farm outlet, publicity, software to manage orders, etc.), as shown by Messmer (2013). In addition, selling through a SFSC may generate a heavier workload (Capt et al. 2011). It takes time to manage or make deliveries, manage and prepare orders, or manage sales points (Lutz and Schachinger 2013). These factors help explain the variable profitability of these chains.

However, logistics is also linked to the ethical dimension and social impacts of short chains. All actors do not pursue the same goals. In the most alternative types of chains the logistics appear to be based more on the creation of strong relationships between actors than on the seek for a strong efficiency of the circulation of product and information (Grauerholz and Owens 2015). In this type of chains actors develop logistics based on direct contacts that not only enable the trade of products but also create interpersonal relationships (Williams et al. 2015) such as in community supported agriculture (CSA). However, this approach is not always possible because farmers don't have enough time and resources and are thus not always able to develop the necessary logistics that would enable direct transactions with customers (Herault-Fournier 2013; Lutz and Schachinger 2013).

Logistics also affect the environmental performance of food supply chains. Life cycle assessments note the importance of production and processing methods (Brodt et al. 2013; Kulak et al. 2015). But logistics (notably, storage and transport) also have a significant impact, as shown by Coley et al. (2009), Meisterling et al. (2009) or Mundler and Rumpus (2012). Simply in terms of transportation from farm to customer, the results can be very different from one chain to another. They can depend on the delivery distances, the load capacity and load rate of a vehicle, the frequency and the organization of deliveries (for example the existence of delivery rounds), and whether the itinerary is urban or rural and the trip solely dedicated to the delivery or to other professional tasks.

Logistics clearly has significant impacts on all chains but the needs vary among them. In Sect. 3.3, we propose a theoretical framework to help understand the diversity of needs and the impacts concerning logistics in SFSCs: the economics of conventions.

3.3 Economics of Conventions: A Framework to Understand Logistics in SFSCs

The theory of conventions provides a relevant grid to understand the diversity of short chains. In this section we will explain why this theory is appropriate for our study, and describe the methodology of our fieldwork.

3.3.1 Diversity as a Key Question in SFSCs

Different theories have been used to help explain the diversity of SFSCs. The economics of proximity has been frequently used to understand the coordination around local products (Chevallier et al. 2014). Depending on the quality of products and the specificity of the resources needed for their production and sale, the needs for geographical and relational proximity are not the same (Pecqueur 2001). Praly et al. (2009) have shown the degree to which this quality has a direct impact on the circulation of products and information throughout the chain. However, this framework does not really consider the competition and power relationships within the chain or among different chains.

By contrast, the actor-network theory, which enables analyses of how differing resources are bundled in food chains, "focuses on the complex negotiations that ensue as these bundles are 'stitched together' and highlights the resultant power relations" (Murdoch 2000: 412). Like the economics of proximity, this theory examines vertical and horizontal coordination and the embeddedness of the activity (Sonnino and Marsden 2006); the latter is a key factor of success (Jarosz 2008). However, as Murdoch et al. (2000), we believe that the theory overly minimizes the question of the quality of products.

In view of these limits, we chose to use the conceptual framework of the economic theory of conventions.

3.3.2 Theoretical Framework: Different Types of Production, as Identified by the Economics of Convention

This framework was applied to short food supply chains by Amemiya et al. (2008) and Touzard (2010) but without consideration of the logistical problematic. The former shows that SFSCs are diverse and based on different types of coordination between actors, particularly because they are based on different conventions in terms of the quality of products and services. The latter looks at French wine cooperatives and shows that changes regarding grape classification conventions and those regarding targeted demand both have impacts on how the wine industry is organized locally and how the involved interactions differ with respect to the surrounding environment. However, even if these studies address the issue of coordination, they do not examine how the diversity of needs for this coordination impacts the management of goods and information flows in supply chains.

Nevertheless, these studies confirm that "conventions" appear to be particularly suited to an analysis of the diversity of logistical organisation because they allow us to identify and evaluate a variety of relationships between actors (Orléan 2004). This diversity is an outcome of the combination of the individual goals of the actors and the context in which they are placed (Boltanski and Thévenot 1991). The links between the two are determined by the performance agreements that make it possible for the actors to understand each other. These agreements may relate to the quality of food products (Eymard-Duvernay 1989; Bénézech 2007).

Salais and Stroper (1993) distinguish four "worlds of production" whose main characteristics are outlined in Table 3.1. In each world, different quality of goods are produced and sold from the most standardized (industrial world) to the least standardized (professional world). In each, the object and degree of uncertainty is different. The degree and the elements of competition are also different. These factors help explain the various amounts of coordination required in each world of production and, therefore, the need for the circulation of product and information (Table 3.1).

On the basis of this framework, we can consider – much like Blanquart and Burmeister (2009), who have studied other types of supply chains – that the elements in Table 3.1 have an impact on the coordination among actors in the supply chain and that this need for coordination influences the management of goods and information flows. Therefore, it also influences logistical needs and their economic, environmental and social impacts.

3.3.3 Fieldwork: 16 Interviews to Analyse the Diversity of SFSCs

Our analysis is based on 16 interviews. To study coordination in the chains, we interviewed producers and retailers who work together. On this basis we created monographs that describe eight different supply chains, each with a retail intermediary acting between the producer and the consumer (Table 3.2). Most producers are farmers, but two are small firms (12 employees each) who place an emphasis on local sourcing.

Producers and intermediaries are both located in the Nord – Pas-de-Calais, a region located in the north of France near the border with Belgium. Most of the intermediaries (retailers) are located in the city of Lille, the most important urban area of the region, with more than 1.1 million inhabitants. The map below (Fig. 3.1) shows the location of producers and intermediaries and the configuration of urban areas in the region.

Table 3.1 The possible worlds of production

	Flexible	Industrial	Professional	Non-material
Types of products	Mass retail products targeting some market segments	Mass retail products	Non-industrial products tailored to meet a specific demand	Material goods suited for mass consumption accompanied by innovative services that firms want to offer
Evaluation of quality	Mass consumption sector standards modified for some clients	Mass consumption sector standards	No standards. Perceived quality shown by price	Constructed in the course of discussions. Importance of rules and methods of production, ethical dimension
Form of uncertainty	Ignorance about demand in immediate future, forecasting difficult	Predictable risk of short-term variation in demand	Uncertainty regarding business partner	Uncertainty regarding the destination market and possibilities for developing activity
Treatment of uncertainty	Products immediately and permanently available	Short and medium-term forecasting of events and behaviours	Process for understanding the capacities and needs of the business partner	Gradual creation of mutual trust
Sphere of competition	Price and quality of product	Price of product	Quality of product	Ability to learn about market and products

Source: Adapted from Salais and Storper (1993)

Table 3.2 The chains studied

Product	Intermediary	Type of SFSC	World of production
Vegetables	Local chain stores	Retailer specialized in fresh products	Flexible
Beer (SME)	Local retailer	E-commerce	
Apple	Chain stores	Supermarkets	Industrial
Dairy products	Local shop	Retailer specialized in dairy products	Professional
Cheese (SME)	Local shop	Shop specialized in local products	Non-material
Vegetables	Local retailer	Box-schemes	
Vegetables	Local farm	Farm outlet and markets	
Dairy products and meat	Local farm	Farm outlet and markets	

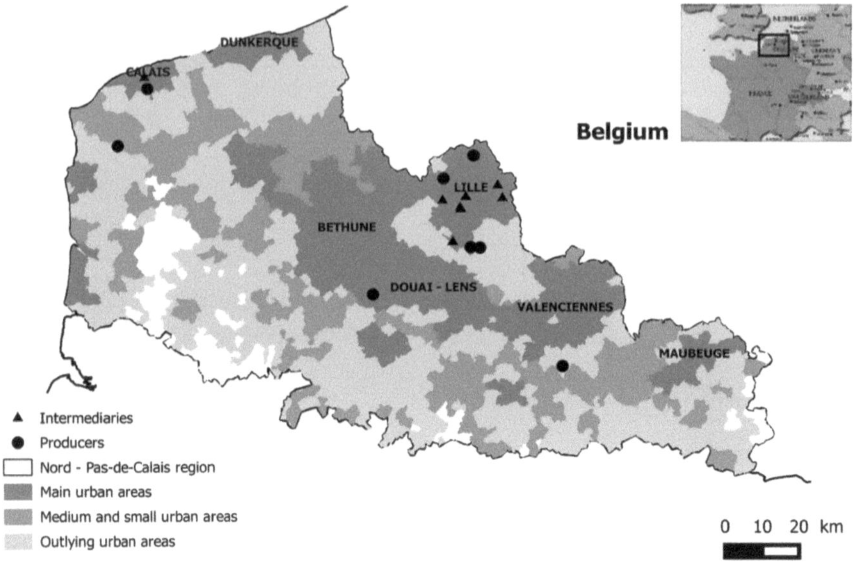

Fig. 3.1 Location of the case studies (Source: Authors, based on INSEE datas and SIGALE® Nord – Pas-de-Calais)

The figure also shows the extent of urbanization in this region, which explains the elevated population density (327 inhabitants/km^2, compared to the national average of 118 inhabitants/km^2). Urbanization and population density can lead to traffic congestion and generate problems of logistics and transportation.[1] These characteristics make this region an appropriate choice for our study. The second reason for this choice is the significant number of short chains operating in the region. They involve 19% of the farms in Nord – Pas-de-Calais, compared to a national average of 15%.

We conducted semi-structured interviews, either face-to-face or by telephone. The retail channels were chosen randomly, the only obligation being that they were each distinct entities supplying end customers in major urban areas of the region. These interviews provide precise information regarding logistics and links between the producer and the intermediary. For each specific chain studied, we asked the two actors to provide information regarding the characteristics of products and clientele, the logistic scheme and the types of relationship between the two actors.

[1] Source: INSEE (2015) Nord-Pas-de-Calais et Picardie: l'influence des grandes aires urbaines. Insee Flash Nord – Pas-de-Calais, 5. http://www.insee.fr/fr/insee_regions/nord-pas-de-calais/themes/insee-flash/IF15_05/IF15_05.pdf. Accessed 7 December 2015.

3.4 Results: Some Chains Have Strong Potential for Sustainability

The four worlds of production are all represented in our sample, although most of the short supply chains studied correspond to the non-material world.

3.4.1 Each World Has a Different Need with Respect to Transportation of Goods

First we analysed how transportation is organized. This analysis shows that certain worlds are much more transport intensive than others.

Two supply chains in our sample correspond to the flexible world: an e-commerce business and a market gardener who supplies shops that specialise in fresh food. The second links a market gardener in the metropolitan area of Lille and a local food chain store with one shop and a logistics hub in the city of Lille. The market gardener delivers several types of vegetables, some that are highly perishable, such as lettuce. He delivers directly to the retailer's hub once or more per day. The goods are then delivered to the sales points which receive them less than 24 h after an order is placed. A just-in-time type of management is firmly entrenched, not only because of the perishable nature of the goods but also to control the quantities that are delivered as precisely as possible. The fact that small additional deliveries may be made during the day reflects a commitment by the intermediary to avoiding gaps on the shelves.

The participants in the e-commerce supply chain are farther apart, and just-in-time management is less present because the producer generally makes deliveries once per month. However, for services to the end customer, just-in-time practices are very much present because the intermediary insists on the need to deliver the product within 24 to 48 h after it has been ordered on the website.

One chain is in the industrial world and typifies those that serve the mass-market food sector. The intermediary is a mass-market retail firm mainly specialised in food. The producer is a group of fruit growers (mainly apples) that supplies the intermediary's local hub (in Lille), and engages a transport service provider. The retailer's central services at the local hub organise orders and deliveries to the points of sales. A minimum of 3 days elapse from the time the producer receives the order to its arrival in the shop, with an intermediate stop at the hub for consolidation. The quantities ordered may vary from one week to the next, based on the sales forecasts made by the central purchasing department.

The professional world is also represented in our sample; one chain operates between a dairy farm and a specialised retailer located in Lille. The latter receives weekly deliveries from the producer, who subcontracts the transportation to a neighbouring farmer. The delivery is made one to 3 days after the order. Although the producer attempts to ensure the best possible availability of products, each party agrees that the nature of certain products (high value added and artisanal products) may occasionally affect that availability. Therefore, the demands related to the flow of physical goods are not as great in this case. But this does not mean that transport is unimportant. The punctuality of deliveries is particularly important from the intermediary's perspective in order to have the product ready for the shop opening.

Four short chains in our sample are in the non-material world. These chains involve sales that could be described as innovative: the sale of vegetable baskets on the Internet, a shop that sells strictly local products at a both a physical location and on the Internet, and two shops selling direct from farms (one that is part of an informal interregional network of producers and the second that sells vegetable baskets and specializes in the sale of a local type of salad). Transportation logistics vary from one chain to another but they also have common features. Most deliveries are weekly and made by the producers themselves, with the exception of one chain where the intermediary provides the transportation. In this case the quantities are small – a few kilos or dozen of kilos per intermediary – and several delivery rounds are organized to deliver the day after an order is placed.

The supply chains in the flexible world appear to be the most transport-intensive because of the predominance of just-in-time, followed by those in the industrial world and the non-material world. Volumes are important in the industrial world chains, but transportation is usually consolidated. The least transport-intensive is the chain in the professional world.

3.4.2 The Organization of Transport: The Importance of Coordination and Its Determinant Factors

Each world is associated with specific types of coordination among actors. This coordination is characterized by the contacts among actors and the information exchanged. The frequency and nature of the contacts depends on the type of products, the uncertainty and the competition in each world.

In the flexible world chains, the producer and intermediary actors described frequent exchanges, but these exchanges were concerned with a limited quantity of commercial information. The contacts were much more frequent in the first supply chain than in the second (at least daily face-to-face or by telephone in the first supply chain, by telephone or Internet for the second). However, despite these differences, the contacts revolved around the same subjects: making orders, occasionally presenting new products from the producer, and discussing volumes and potential sales. For example, all of the actors confirmed that they had no conversations regarding consumer feedback or the producer's logistical constraints.

In both chains, the logistical capabilities of the producer is an important selection criterion for the intermediary. The actors also expressed the importance of the quality of the product as well as the price. The price concern is particularly apparent in the e-commerce supply chain in which the intermediary works directly with the producers when it is more economically profitable than purchasing from a wholesaler. The logistic scheme is designed to guarantee just-in-time procurement. In the first chain, this is particularly facilitated by the closeness of the producer with the intermediary's hub (approximately 15 km). In this case, just-in-time is also the first goal of the information exchange because there is no uncertainty regarding the high standardized quality of products.

In the chain representing the industrial world, the links between actors are weaker. The producer and intermediary meet at the beginning of the sales campaign and during the year to discuss the anticipated volume of production and sales, and develop mid-term forecasts. These meetings also provide an opportunity to discuss product grading, which addresses the criterion of quality standards applied by this type of mass market distribution channel. There is minimal uncertainty regarding the quality in this case because the product varieties are very common (not always true in the flexible world chains). The production objective here is to satisfy strong demand for products whose origins are not stated at the point of sale (in contrast to the flexible world). Occasional meetings and interactions during ordering constitute the majority of contacts between the two parties. There are some contacts undertaken to plan promotional campaigns, which indicates that competition is essentially based on price. This aspect is further evidenced by the fact that the intermediary modifies the purchase price to the producer on a daily basis, which was not typical in the flexible world chains. Therefore, the most important issue in this industrial world chain is to offer products at a competitive price. This competitiveness is necessary for a very standardized type of product that can be found everywhere else, both for the intermediary selling to the consumer and for the producer selling to the intermediary. Although the actors have discussions regarding volume, the producer has no guarantee of the price that will be paid or the quantity that will be ordered, from one week to the next.

The dynamics are different in the professional world supply chain. The absence of objective standards for quality: label, or the well-known characteristics of a clearly identified variety, is a major source of uncertainty on the product (farm-made yogurts). Therefore, an agreement on quality must be achieved between the producer and the intermediary. In the observed chain, the actors had preliminary meetings and discussions before working together. This was followed by a period of sales tests. In our example, these tests were important for both actors. The intermediary needed to confirm the capabilities of the producer to supply the product, and the producer needed reassurance that the intermediary was able to market the unique nature of a more expensive product to consumers. The goal expressed by the actors of this chain was to establish a long term relationship, offering products that are "out of the ordinary". We observed a commitment by both parties: the producer agrees not to place products in the shops of the intermediary's close competitors and the intermediary agrees not to seek other producers of the same product. None of the actors identified the product's price as a major problem. Although contacts between the actors, primarily face-to-face, were frequent at the beginning of the relationship, the need for them progressively diminished and they had become very rare by the time of the interview.

All the intermediaries involved in a non-material chain attempt to offer a variety of products; but more importantly, they want to offer new types of services. During the interviews, they expressed uncertainty regarding the future development and longevity of their supply chain. They attempt to develop strategies that can be adjusted for changes in demand. This usually means varying the quantities of product offered on the basis of information they have gathered regarding the volatility and seasonality of demand. In the cases studied, the foodstuffs offered are not readily distinguished from those distributed by the mass retail sector. Most contacts between producers and intermediaries occur during weekly restocking (face-to-face) or during the ordering process (usually by telephone). In three of the four cases, the intermediary and producer occasionally meet in person to discuss quality and availability of the products, and information on customer feedback and demand. It is particularly important to adapt product availability to end demand, and assure the reliability of deliveries, when applicable, which are extremely important for developing customer loyalty. The importance of logistics is also evident in the fact that every actor (producer and intermediary) expressed the desire to meet consumer needs in the most effective manner. Trust also appears to be an important element in these new retailing concepts.

The different priorities in each type of world involve different coordination needs, and consequently, different logistic issues that determine the organization of transport and exchanges of information in the supply chains. Therefore, these priorities can have different impacts on the transportation of products in urban areas, and different economic, environmental and social effects.

3.4.3 *Coordination and Transport Needs: Two Elements that Help Us Understand the Impacts of SFSCs in Urban Areas*

To summarize the impacts of each type of chain and logistics, we first considered whether the world is more or less transport-intensive, and if transport's consolidation is possible or not. We consider these elements because intensive transport needs can generate negative environmental (pollution) and economic effects (extra costs for producers). We then examined the types of relationships in these chains to help us understand the social impacts (Table 3.3): How often do actors interact, what information is exchanged, and do these relationships appear to be stable on a long-term basis? This last element also has an impact on economic performance because it is an indicator of the level of competition. A high level can be negative for producers, particularly with regard to prices or when the intermediary is more powerful (because of its size, for example, or the producer's level of dependency on the intermediary).

Table 3.3 Outlines the main characteristics of the four worlds, their logistics and sustainability

	Flexible	Industrial	Professional	Non-material
Supply chain	Supplying a small food retailer	Supplying mainly foodstuffs via a distribution hub that serves the mass distribution sector	Supplying specialized independent shops directly	New modes of foodstuff distribution (e-commerce, shops selling local products)
Priorities	Permanent supply of goods with a good quality-price ratio	Supplying large volumes of products at a low price	Offering unique products	Offering products associated with innovative services
Logistical stakes	Rapid and frequent transport of low volumes	Rapid transport of consolidated flows	Transport of small volumes under conditions that are favourable to the producer and distributor	Rapid transport of small volumes, exchange of non-logistical information
Sustainability	Environmental –	Environmental +	Environmental +	Environmental+/–
	Economic –	Economic –	Economic +	Economic?
	Social –	Social –	Social +	Social +

For the chains in flexible world, being transport-intensive suggests more potential adverse environmental impacts. These chains require producers to make frequent trips to deliver small volumes. From a social standpoint, the relationship between the producer and the intermediary appears to be basically commercial but stable (the actors have work with each other for several years). From an economic perspective, the need to make frequent deliveries also implies elevated transport costs.

The negative environmental impact of the industrial world chains should be smaller because the volumes transported are more consolidated, which is a means of reducing the environmental impact by limiting the number of trips. However, the economic and social impacts of this type of supply chain may be more problematical, in particular for the producer who can suffer from intense price competition, and uncertainty in his relationship with the intermediary.

In contrast, the professional world appears to be very positive from the social and economic perspective because of less price competition and a more balanced producer-intermediary relationship. Although a certain efficiency of the logistics is still required, the need for very rapid and daily deliveries is absent in this chain. This advantage, combined with the nature of the relationship between actors, leads to a more sustainable chain where both parties can work towards the organization of less frequent and more optimized trips.

The environmental impacts of non-material chains are difficult to evaluate. Although the distances are short in our monographs, there may be numerous trips for small volumes. These frequent deliveries or trips for supplies can generate high transport costs from an economic perspective, although they can also have a positive effect because they make it possible for actors to directly exchange important information about demand. The actors in these chains are similar in terms of size, which suggests the possibility of a more balanced relationship. All of the above factors reflect generally positive social effects.

3.5 Conclusion

Many cities promote short supply chains as one of the solutions to provide sustainable food to citizens; however, their logistics performance appears variable. The "worlds of production" provide a framework for understanding these chains and their diversity. Looking at the need for coordination between actors in diverse supply chains helps explain how logistics is organized. Although the impacts of logistics in the non-material world may be positive, they can be difficult to evaluate. The professional world appears promising. In this chain there is a stronger emphasis on product quality and less dependence on logistics as there is less need for rapid and frequent transport of small quantities of goods. In addition, there seem to be more equilibrium in the relationship between the producer and the intermediary, which provides positive social and economic impacts. However, further investigations are needed to confirm these results, notably because the type of product has an impact on the frequency of supply needs and consequently on logistics. It would be important to study "professional" chains, providing other products that are more or less perishable than yogurts.

These conclusions concur with those of Amemiya et al. (2008), although we have supplemented their analysis of the chains by considering the concrete logistical issues that they generate, as did Burmeister (2000), who studied the different needs for circulation of products according to the type of supply chain, with similar conclusions. However, the latter did not look at SFSCs. Nevertheless, these authors have studied very different chains, which makes their classification easier. We investigated the diversity inside one "family" of chains: local food chains with one intermediary. Consequently, the differences between worlds are diminished. The need for rapid transport in three of the four worlds reflects this aspect and the degree of 'porosity' in the worlds of production. This suggests that, beyond the diversity of coordination, there is a partial community of needs regarding logistics or at least transport. In addition, the low number of cases also suggests further investigations.

Nevertheless, this study helps us understand the diversity of short food supply chains and why logistics may be organized differently from one chain to another. The type of product and its quality, the uncertainty that actors must confront, and the level and "sphere" of competition help explain the variety of supply chains, their complexity, their priorities and their logistic issues, particularly with regard to transport. All of these elements have an influence on the sustainability of these chains and help us understand why they can also have adverse impacts. Thus, improving the ability of short food supply chains to provide sustainable food systems for cities is not simply a technical or technological challenge to find one-size-fits-all solutions to reduce transport and its impacts. It requires a more global examination of the complexity of these determining factors.

References

Amemiya H, Bénézech D, Renault M (2008) Les circuits courts: un « monde de commercialisation » interpersonnel? In: Maréchal G (Coord.) Les circuits courts alimentaires. Bien manger dans les territoires. Educagri, Dijon, pp 113–125

Aubry C, Chiffoleau Y (2009) Le développement des circuits courts et l'agriculture périurbaine: histoire, évolution en cours et questions actuelles. Innov Agronomiques 5:53–67

Aubry C, Kebir L (2013) Shortening food supply chains: a means for maintaining agriculture close to urban areas? the case of the French metropolitan area of Paris. Food Policy 41:85–93. https://doi.org/10.1016/j.foodpol.2013.04.006

Bénézech D (2007) Les ressorts de l'échange de produits alimentaires différenciés: de l'information à la confiance. In: Amemiya H (ed) L'agriculture participative. Presses Universitaires de Rennes, Rennes, pp 137–149

Blanquart C, Burmeister A (2009) Evaluating the performance of freight transport: a service approach. Eur Transp Res Rev 1(3):135–145. https://doi.org/10.1007/s12544-009-0014-5

Blanquart C, Gonçalves A, Kebir L, Petit C, Traversac JB, Vandenbossche L (2010) The logistic leverages of short food supply chains performance in terms of sustainability. In: 12th world conference on transport research. http://www.civil.ist.utl.pt/wctr12_lisboa/int_04_program_conference.htm

Boltanski L, Thévenot L (1991) De la justification. Les économies de la grandeur. Gallimard, Paris

Bosona T, Gebrensenbet G, Nordmark I, Ljuinberg D (2011) Integrated logistics network for the supply chain of locally produced food. Part II: Location and route optimization analysis. J Serv Sci Manag 2011(4):174–183. https://doi.org/10.4236/jssm.2011.42021

Bosona T, Nordmark I, Gebrensenbet G, Ljuinberg D (2013) GIS-based analysis of integrated food distribution network in local food supply chain. Int J Bus Manag 8(17):13–34. https://doi.org/10.5539/ijbm.v8n17p13.

Brodt S, Kramer KJ, Kendall C, Feenstra G (2013) Comparing environmental impacts of regional and national-scale food supply chains: a case study of processed tomatoes. Food Policy 42:106–114. https://doi.org/10.1016/j.foodpol.2013.07.004

Burmeister A (2000) Familles logistiques. Propositions pour une typologie des produits transportés pour analyser les évolutions en matière d'organisation des transports et de la logistique. Rapport pour le PREDIT. http://urbamet.documentation.developpement-durable.gouv.fr/documents/EQUTEXA0001916/EQUTEXA0001916.pdf

Capt D, Chiffoleau Y, Gauche A (2011) Elaboration d'un référentiel technico-économique dans le domaine des circuits courts de commercialisation. Partie 1: Cadre d'analyse, méthodologie et synthèse des résultats Légumes et Produits laitiers en circuits courts. Ministère de l'Agriculture, de l'Alimentation, de la Pêche, de la Ruralité et de l'Aménagement du Territoire. http://prodinra.inra.fr/?locale=fr#!ConsultNotice:162944

Chevallier M, Dellier J, Plumecocq G, Richard F (2014) Dynamiques et structuration des circuits courts agroalimentaires en Limousin: distance institutionnelle, proximités spatiale et relationnelle. Géographie, Économie, Société 3(16):339–362

Cleveland DA, Müller NM, Tranovich AC, Mazaroli DN, Hinson K (2014) Local food hubs for alternative food systems: a case study from Santa Barbara County, California. J Rural Stud 35:26–36. https://doi.org/10.1016/j.jrurstud.2014.03.008

Coley D, Howard M, Winter M (2009) Local food, food miles and carbon emissions: a comparison of farm shop and mass distribution approaches. Food Policy 34(2):150–155. https://doi.org/10.1016/j.foodpol.2008.11.001
Deverre C, Lamine C (2010) Les systèmes agroalimentaires alternatifs. Une revue de travaux anglophones en sciences sociales. Econ Rural 3(317):57–73
European Commission (2014) Short food supply chains. http://ec.europa.eu/agriculture/organic/consumer-trust/certification-and-confidence/short-supply-chain/index_en.htm
Eymard-Duvernay F (1989) Conventions de qualité et formes de coordination. Revue Economique 40(2):329–360
Feagan R (2007) The place of food: mapping out the 'local' in local food systems. Prog Hum Geogr 31(1):23–42. https://doi.org/10.1177/0309132507073527
Grauerholz S, Owens N (2015) Alternative food movements. In: Smelser NJ, Baltes PB (eds) International encyclopedia of the social & behavioral sciences. Elsevier, Amsterdam, pp 566–572
Hayden J, Buck D (2012) Doing community supported agriculture: tactile space, affect and effects of membership. Geoforum 43(2):332–341. https://doi.org/10.1016/j.geoforum.2011.08.003
Herault-Fournier C (2013) Est-on vraiment proche en vente directe ? Typologie des consommateurs en fonction de la proximité perçue dans trois formes de vente: AMAP, Points de vente collectifs et Marchés. Manag Avenir 64(6):167–184. https://doi.org/10.3917/mav.064.0167
Hinrichs C (2003) The practice and politics of food system localization. J Rural Stud 19(1):33–45. https://doi.org/10.1016/S0743-0167(02)00040-2
Jarosz L (2008) The city in the country: growing alternative food networks in metropolitan areas. J Rural Stud 24(3):231–244. https://doi.org/10.1016/j.jrurstud.2007.10.002
Kulak M, Nemeceka T, Frossard E, Chable V, Gaillard G (2015) Life cycle assessment of bread from several alternative food networks in Europe. J Clean Prod 90:104–113. https://doi.org/10.1016/j.jclepro.2014.10.060
Lendrevie J, Levy J, Lindon D (eds) (2009) Mercator: Théorie et pratique du marketing. Dunod, Paris
Ljungberg D, Jüriado R, Gebresenbet G (2013) Conceptual model for improving local food supply chain logistics. http://www2.wctr2013rio.com/general-proceedings/documents/1451.pdf
Lutz J, Schachinger J (2013) Do local food networks Foster socio-ecological transitions towards food sovereignty? Learning from real place experiences. Sustainability 2013(5):4778–4796. https://doi.org/10.3390/su5114778
Martikainen A, Niemi P, Pekkanen P (2014) Developing a service offering for a logistical service provider—case of local food supply chain. Int J Prod Econ 157:318–326
Meisterling K, Samaras C, Schweizer V (2009) Decisions to reduce greenhouse gases from agriculture and product transport: LCA case study of organic and conventional wheat. J Clean Prod 17(2):222–230
Messmer JG (2013) Les circuits courts multi-acteurs: emergences d'organisations innovantes dans les filières courtes alimentaires. Rapport de recherche INRA. http://www7.inra.fr/lecourrier/wp-content/uploads/2014/11/Circuits-courts-Messmer-2013.pdf
Morganti E, Gonzalez-Feliu J (2015) City logistics for perishable products. The case of the Parma food hub. Case Stud Trans Policy 3(2):120–128. https://doi.org/10.1016/j.cstp.2014.08.003
Mundler P, Rumpus L (2012) The energy efficiency of local food systems: a comparison between different modes of distribution. Food Policy 37:609–615. https://doi.org/10.1016/j.foodpol.2012.07.006
Murdoch J (2000) Networks — a new paradigm of rural development? J Rural Stud 16(4):407–419. https://doi.org/10.1016/S0743-0167(00)00022-X
Murdoch J, Marsden T, Banks L (2000) Quality, nature, and embeddedness: some theoretical, considerations in the context of the food sector. Econ Geogr 76(2):107–125. https://doi.org/10.1111/j.1944-8287.2000.tb00136.x.

Nordmark I, Ljuinberg D, Gebrensenbet G, Bosona T, Jüriado R (2012) Integrated logistics network for the supply chain of locally produced food. Part II: Assessment of E-trade, economic benefit and environmental impact. J Serv Sci Manag 2012(5):249–262
Orléan A (2004) L'Economie des Conventions: définitions et résultats. In: Orléan A (dir.) Analyse économique des conventions. PUF, Paris, pp 9–48
Parker G (2005) Sustainable food? Teikei, Co-operatives and food citizenship in Japan and the UK. Working papers in Real Estate & Planning, 11/05
Pecqueur B (2001) Qualité et développement territorial: l'hypothèse du panier de biens et de services territorialisés. Econ Rural 261:37–49
Praly C, Chazoule C, Delfosse C, Bon N, Cornee M (2009) La nation de « proximité » pour analyser les circuits courts. http://www.reseaurural.fr/files/u1/asrdlf_praly_et_al_2009.pdf
Salais R, Storper M (1993) Les mondes de production. Enquête sur l'identité économique de la France. Editions de l'Ecole des Hautes Etudes en Sciences Sociales, Paris
Savy M (2007) Le transport de marchandises. Editions d'Organisation, Groupe Eyrolles, Paris
Schliwa G, Armitage R, Aziz S, Evans J, Rhoades J (2015) Sustainable city logistics — making cargo cycles viable for urban freight transport. Res Transp Bus Manag 15:50–57. https://doi.org/10.1016/j.rtbm.2015.02.001
Sonnino R, Marsden T (2006) Beyond the divide: rethinking relationships between alternative and conventional food networks in Europe. J Econ Geogr 6(2):181–199. https://doi.org/10.1093/jeg/lbi006
Touzard JM (2010) Ancrage territorial et construction de règles dans des coopératives viticoles du Midi. In: Muchnik J, Sainte-Marie C (coord.) Le temps des SYAL. Editions Quae, Versailles, pp 265–279
Viljoen A, Wiskerke JSC (2012) Sustainable food planning: evolving theory and practice. Wageningen Academic Publishers, Wageningen
Williams LT, Germov J, Fuller S, Freij M (2015) A taste of ethical consumption at a slow food festival. Appetite 91:321–328. https://doi.org/10.1016/j.appet.2015.04.066

Chapter 4
Interactions Between Agriculture and the City: A Systemic Approach to Examine Sustainability in Meknes (Morocco)

Jean-Pierre Chery, Elodie Valette, Patrick Dugué, Pascale Philifert, and Abdellaoui El Hassane

Abstract A growing population, urbanisation, and the economics of agricultural development are among the principal issues that Morocco faces. These issues come with important consequences, such as a strong demand for housing and changes in land tenure. National public policies address these issues sectorally through regional and local actions. This research project examines the relationships between questions of environment, urban development, sustainable agriculture, and local and national logics. We studied these relationships through a systemic representation aimed at a better understanding of the operative elements, as well as the state of individual and collective action in Meknes. As the sixth largest urban centre in Morocco, Meknes illustrates the difficulties of merging growth in agricultural production and urbanisation. Urban planning in the region has favoured the development

J.-P. Chery (✉)
TETIS, AgroParisTech, CIRAD, CNRS, IRSTEA, University of Montpellier, Montpellier, France
e-mail: jean-pierre.chery@agroparistech.fr

E. Valette
ART-DEV, University of Montpellier, CIRAD, CNRS, University of Paul Valery, University of Perpignan, Montpellier, France
e-mail: elodie.valette@cirad.fr

P. Dugué
CIRAD, UMR Innovation, Innovation, University of Montpellier, CIRAD, INRA, Montpellier Supagro, Montpellier, France
e-mail: patrick.dugue@cirad.fr

P. Philifert
Université Paris Ouest Nanterre, UMR LAVUE, Nanterre, France
e-mail: pascale.philifert@u-paris10.fr

A. El Hassane
Ecole Nationale d'Agriculture de Meknès, Meknès, Morocco
e-mail: elhassanea@aol.com

C.-T. Soulard et al. (eds.), *Toward Sustainable Relations Between Agriculture and the City*, Urban Agriculture, https://doi.org/10.1007/978-3-319-71037-2_4

of real estate and the application of a law designed to convert collective lands to private ownership, creating significant changes in land use, often at the expense of agriculture. A systemic approach helps us identify the externalities that influence public policies and local actions, and underlines the need for a multi-level examination of the relationships between cities and agriculture.

Keywords Land tenure · Public policy · Urban planning · Sustainabilty

4.1 Introduction

In the last 15 years, numerous studies and examples have shown the advantages of integrating urban and agricultural questions in developing countries. Alarming forecasts concerning growth in urban demographics (Hove et al. 2013), climate change, and specific events such as the hunger riots during the 2007–2008 famine (Berazneva et Lee 2013) have prompted researchers and policy makers to reconsider the value of urban and periurban agriculture. Today, these forms of agriculture are increasingly included in the search for strategies to improve food security for growing urban populations.

Although urban and periurban agriculture have always existed and provided various social functions regarding food production and the environment (Temple, etc. 2004), their recognition and support through public policies are still emerging in many countries. The position of public authorities regarding these forms of agriculture varies from rejection to full support. Urban agriculture, particularly when situated in intra-urban interstices, is sometimes discouraged, as in Harare, Zimbabwe (Mbiba 2000), and Yaoundé, Cameroun (Yemmafouo 2014), where it is considered incompatible with a modern vision of urban development. However, in Ghana's capital city, Accra, where market food supplies are precarious, the municipality has implemented measures protecting food production in the city, and actually reserved land for agricultural activities. Ghana is also home to the operation, Feed Yourself, initiated in 1972 as a means of encouraging household agricultural activities among urban residents (Obosu-Mensah 2002).

Pearson et al. (2010) underline the need to integrate agricultural issues in urban development policies aimed at preserving landscapes and the environment at various scales. Analysing sustainability demands that we question existing logics and actions from a perspective that anticipates the evolution of agriculture's place in urban and periurban areas. Agricultural land has been, and often remains, a consumable resource for urban development with little evaluation of the social, economic and environmental impacts.

In this chapter we use a transverse analysis of the relationships between agriculture and the city of Meknes, Morocco, to present an example of the difficulties of integrating agricultural and urban problematics.

A large percentage (42.6%) of Morocco's population is considered rural, and 24% of the working population was employed in agriculture in 2012 (FAOSTAT 2015). But the growing urban share of the Moroccan population (+2.1% between 2004 and 2014) has amplified urbanization pressures on land use. Projections estimate that by 2025, some 90,000 ha of agricultural land will be consumed by urban development at the rate of approximately 4500 ha per year (INAU 2005). Meknes has experienced significant urban growth: Its population rose from 494,100 in 1994 to 717,594 in 2014 (Haut-Commissariat au Plan 2015), and between 2001 and 2011, the built surface area increased from 3084 ha to 4087 ha, most often at the expense of agricultural lands (Valette et al. 2013).

For the most part, Moroccan actors still view urban and agricultural issues through distinctly separate perspectives that fail to address the value of relationships between cities and agriculture. On one hand, the government has a long history of defining agriculture as a national priority, and supports this position with an agricultural strategy launched in 2008 known as the Green Morocco Plan (Plan Maroc Vert, in Ministère de l'Agriculture 2008; Akesbi 2012). But on the other hand, housing has also been a priority, leading to the implementation of mechanisms that directly or indirectly favour development over agriculture in urban and periurban land-use decisions.

In this chapter we use the case of Meknes to examine this divergence in perspectives. We begin by revealing the systemic approach of our research, and then identifying the principal elements that characterize the relationships between this city and agriculture. Finally, we offer a graphic representation of the system of relationships between the city of Meknes and agriculture, and a concluding discussion on their relevance to the issues of sustainability in urban agriculture.

4.2 Using Feedback Loops to Study Interactions Between Cities and Agriculture

Here we present the general concepts of the systems analysis in order to define the interpretative framework that we use in the study. This approach through a systems analysis has long been used as a powerful conceptual framework for questioning economic and social issues associated with the environment (Meadows et al. 1972). It has been an integral tool of research during the emergence and diffusion of the concept of sustainable development (WCED 1987; Hodge 1996), and in the analysis of sustainability in numerous systems.

Nevertheless this tool has not often been used at the regional and local geographical scales which also harbour many of the issues of sustainability (Voiron-Canicio 2005). We hypothesize that a system of interactions between cities and agriculture exists (Valette et al. 2012): This system is created by the dynamics of integration when cities and agriculture develop interactions that reinforce their complementary aspects or mutual dependence, or by contrast, the dynamics of rupture when they are in competition, such as for the same resources. The complexity of relations at the heart of this system points to the relevancy of a systems analysis which offers a methodology enabling the identification of the numerous interactions and various elements as well as their dynamics.

4.2.1 A system's Analysis Approach: The Interactions between Cities and Agriculture Are Complex and Reciprocal

A systems analysis is characterized by four principal concepts (Durand 2006): interaction, globality, organization, and complexity. The system's qualitative representation is a model based on knowledge acquired and assembled by the modeller in an effort to express the studied system's characteristics as they relate to these four concepts. This constructivist style of qualitative modelling is generally accomplished with graphics capable of presenting the entire ensemble of system elements and their interactions. The presentation of these concepts here uses simple thematic examples of the system of interactions between the city and agriculture to highlight their contribution to the question of sustainability, before the application of the analysis to the case of Meknes that follows.

Interactions Between Cities and Agriculture

The concept of interaction addresses the causal links in a system. The reciprocal influence between the elements represented in these links establishes a circular causality which the systemic approach identifies as feedback loops.

In a system of interactions between a city and agriculture, feedback loops can concern energy (food as a source of calories for the urban population), materials (the operations concerning the physical manipulation of agricultural products and by-products), or information (the interactions between the land use demands of urban expansion, landscaping, and agricultural production). The recognition of these feedback loops is essential in the study of sustainability, where the exploitation of resources assures the development and maintenance of an entity, and the depletion of those resources negatively impacts the sustainability of the entity.

Globality: A Multidisciplinary Approach to the System of Interactions Between Cities and Agriculture

The concept of globality is essential in a systemic approach. Suggesting the existence of a system of interactions between cities and agriculture means hypothesizing that the system has aspects that cannot be reduced to the sum of urban and agricultural sub-systems. Figure 4.1 represents causal chains in different urban and agricultural examples (cases a and b), and a more global depiction of the system (case c), where the impacts of food needs on agricultural activities and the impacts of agricultural lands on the city are merged and considered together.

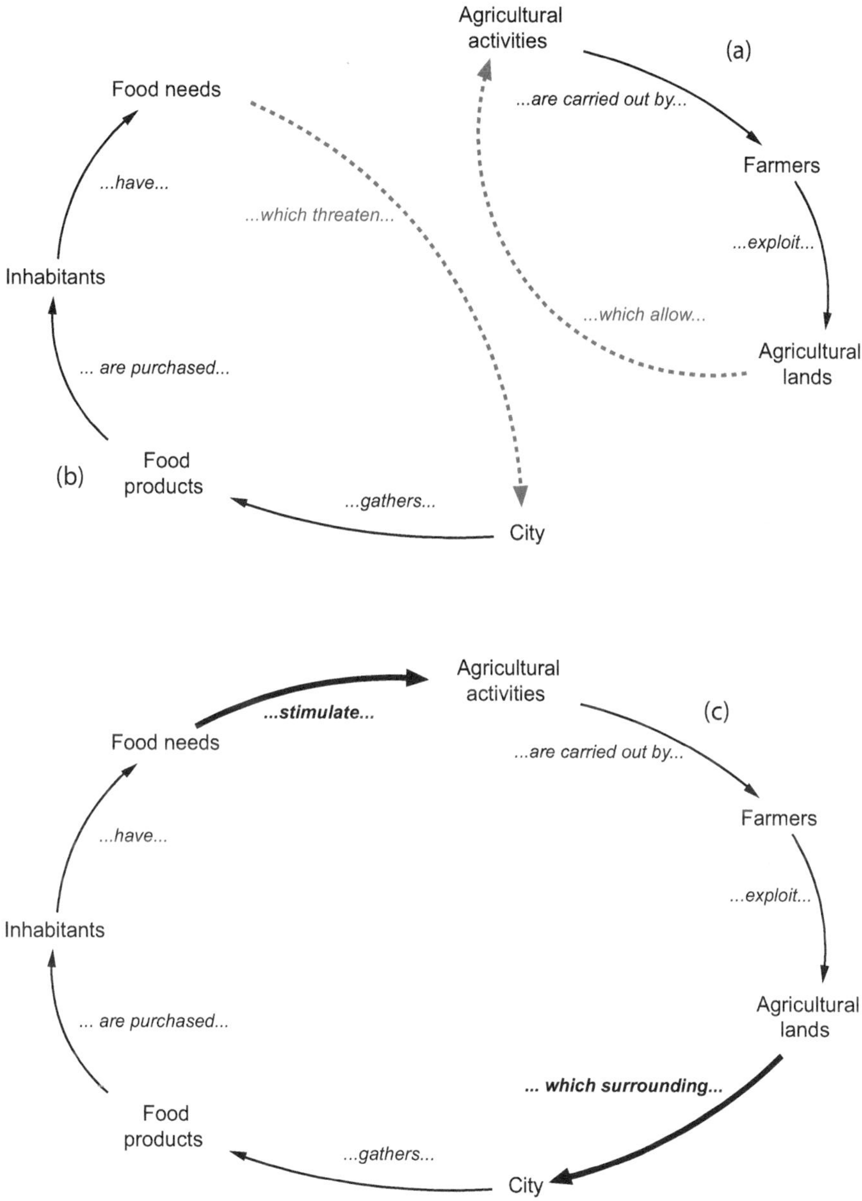

Fig. 4.1 Globality of the system of interactions between cities and agriculture: (**a** and **b**) example of the separate elements of two systems, and (**c**) elements associated through a causal relationship

The concept of globality helps us understand the relationships between the three pillars of sustainability. The globality of the system of interactions between cities and agriculture can be examined from this perspective through (i) the different actors (inhabitants, decision-makers, etc.) and their social status and strategies, (ii) the economic factors and entities (employment, agricultural production, market characteristics, etc.), and (iii) through the environmental dimensions (state of natural spaces, biodiversity, pollution, etc.).

Organisation: Does the System of Interactions Have a Goal?

Analysing the organization allows us to consider both the structure, and the functioning of a system. This dynamic focuses the attention on actors' strategies. Figure 4.2 illustrates the same type of causal structure as seen in Fig. 4.1c, but it is based on farmers' strategies for using periurban agricultural land for export production.

This example illustrates the rise in food prices as a consequence of farmers' strategies and the demand from international markets for certain products. Furthermore, analysing the system's organization allows us to consider its relationship with other systems. Farmers' strategies are not only defined by the needs of local consumers, but also by external markets in the context of globalization. The import price of foodstuffs is equally subject to factors independent from the local context. This exposure to a global system (worldwide) affects the organization and in particular, the functioning of the system of interactions between cities and agriculture.

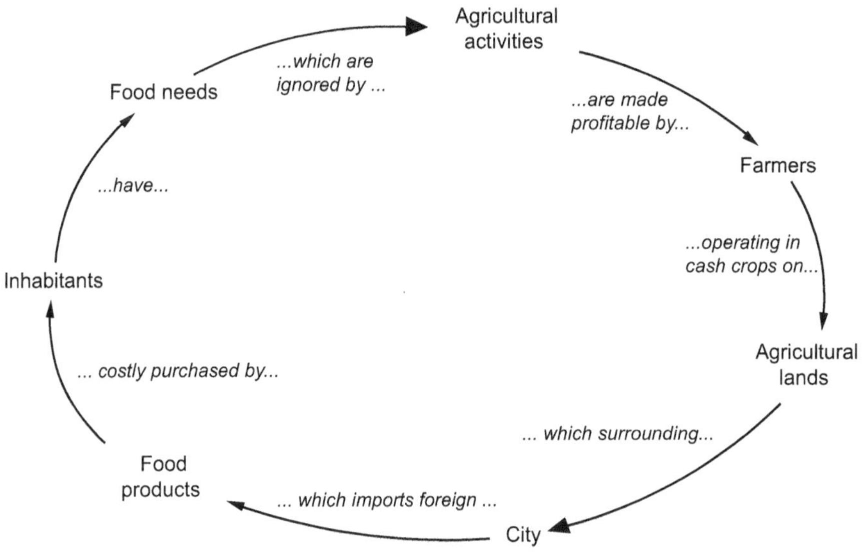

Fig. 4.2 The same structure applied to a different function from the system of interactions between cities and agriculture

A system is sustainable when it maintains or regenerates the elements that compose it. Assuming that nothing changes at the core of a system's structure, the organizational analysis helps us assess the long-term sustainability of the system's functions. Projecting the future state of a system also requires an evaluation of how it has functioned in the past. The process of forecasting sustainability uses this knowledge to elaborate a scenario as a reference for 'business as usual' in possible futures.

System Complexity: How Do We Question the Possible Course of Relationships Between Cities and Agriculture?

The fourth concept of a systemic approach suggests that a complex system has the capacity to change its own organization (auto-organization) and allow the rise of new system properties (emergence) such as resilience. While analysing the organization helps us view the system from a synchronic perspective, examining the complexity delineates the system dynamically and diachronically, exposing its evolution, adaptation and resistance (Ambrosio-Albalá and Bastiaensen 2010).

In the system of interactions between a city and agriculture, these evolutions can be illustrated by the changes in land use. Figure 4.3 shows the impact on farmland from changes in land use during a succession of three political periods in Morocco: pre-colonial, colonial, and post-colonial. The colonial period was marked by the monopolization of agronomically valuable land and a rupture from the previous, pre-colonial, organization of land-use and ownership. In post-colonial Morocco, recovered agricultural land was organized in collectives as part of an attempt to direct the modernization of agriculture under the control of a central authority.

The capacity of a system to organize itself allows it to evolve. Events such as internal innovation or external interventions stimulate the reorganization of relationships

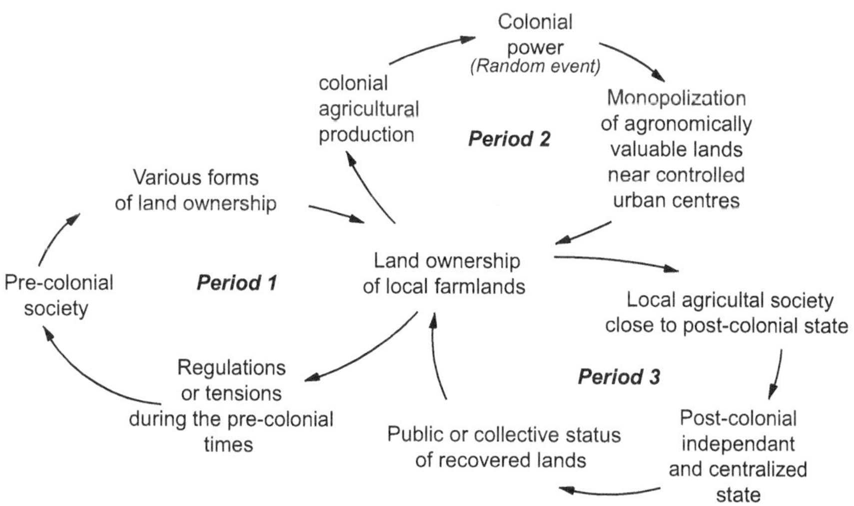

Fig. 4.3 An example of the systematization of periurban land-use: the vagaries of Morocco's colonial period, and its consequences

between resources present in a system of interactions between a city and agriculture. For example, this restructuring can entail social or economic organization, or the variety of natural and exploitable goods. The establishment of sustainability in the system generally requires researching and favouring events most likely to provoke change. A complex system such as a regional or local system of land-use, which is often influenced by national and global perspectives, requires an analysis of multiple dimensions in order to evaluate the organizational aspects that optimise this evolution towards sustainability.

4.2.2 The Feedback Loops that Affect the Complex System of Interactions Between a City and Agriculture

The system's feedback loops act as regulators or amplifiers. As regulators, they compensate for the actions of one or several phenomena. The example of the exploitation of water for agricultural irrigation provides an illustration. The water users control the reserves and can regulate irrigation usage to allow the regeneration of the source through its normal cycle. Figure 4.4 illustrates this balancing loop; a positive sign (+) is associated with the causal arrow between the level of water reserves and the water used for irrigation, indicating that variations occur in the same direction (when the water reserve increases the irrigation can increase or when the reserve decreases the irrigation must decrease). A negative sign (−) is associated with the causal arrow depicting the inverse interaction of the relationship and its opposing consequences; this causal arrow shows that increases in irrigation have the opposing effect of decreasing reserves. This balancing loop (indicated by a negative sign in the center) is established in the relationship between two opposing causalities: an increase provokes a decrease for each element in the loop.

The amplifying feedback loops are either dubbed "negative circles" in contexts where the evolutions are detrimental, or "positive circles" when evolutions are beneficial. The city's increasing population is an important element in the development of amplifying feedback loops: for example, the demographic growth requires an increase in the number of wells to meet the requirements in volume for urban drinking water. When the distribution of additional water is secured, the population can continue to grow which once again increases the need for water resources (Fig. 4.5).

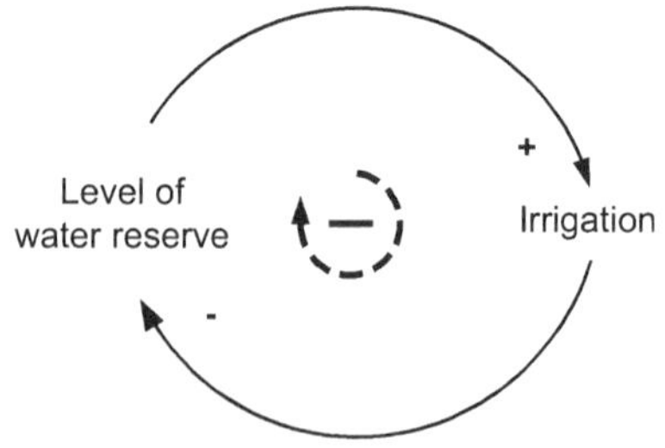

Fig. 4.4 A balancing feedback loop: water reserves and irrigation

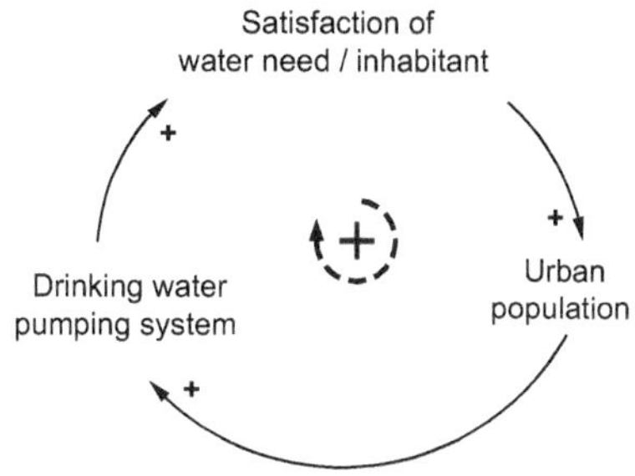

Fig. 4.5 A reinforcing feedback loop: water resources and the urban population

Inversely, a lack of information, or delay in the recognition of phenomena that are not integrated in the organization of the system can modify the state of a resource and establish a negative circle. In the example of water usage, the exploitation of the water table creates competition among urban and agricultural actors. In addition to the volume of usage, city waste water affects the quality of the groundwater, thus negatively affecting resources for irrigation. Without the emergence of new regulations that integrate the urban actor into the system, the water resource declines in both quantity and quality.

When studying an open system that is also a given area, it is more difficult to define the ensemble of elements and interactions, particularly external elements and their influence. Issues with the hierarchies, relationships, and interactions with neighbouring areas affect the interpretation of the dynamics of the system and the evaluation of sustainability.

4.3 The System of Interactions Between the City of Meknes and Agriculture

The interactions between the city of Meknes and agriculture are well suited to a systemic characterization, and the circular, causal relationships involved in these interactions point to dynamics that are clearly sectoral. Different studies conducted in Meknes through the research program of ANR DAUME (2011–2015) define terrains and furnish analyses of the relationships between urban and agricultural sectors of Meknes (François et al. 2013; Valette et al. 2013; Debolini et al. 2015; Dugué et al. 2015) within the larger Moroccan context (Philifert 2011; Valette et Philifert 2014). Using these studies as a base, we have examined the evolution of the region's agriculture from the perspective of its quantifiable dimensions (such as surface areas) and from the perspective of the diversity and organization of the actors (social actors either individual or collective, their strategies, their regulatory methods, their competition for resources), and the objects concerned (the information, resources, revenue, power).

4.3.1 The Meknes Region in the Larger Moroccan Context

The Meknes region can be considered as generally representative of the Moroccan issues of urban growth and the modernization of agricultural production. These issues are characteristic in both the local and national contexts.

The region is divided into four urban areas, Meknes, Ouislane, Toulal, and Boufekrane, and six rural areas, Aït Oualal, Dar Oum Soltan, Dkhissa, Oued Jdida, Sidi Slimane Moul Kifane, and Majjate (Fig. 4.6). The total surface area is approximately 590 km^2.

In 2010, Meknes was the sixth largest metropolitan region in Morocco (Gazel et al. 2011). Average annual growth was +1.95% between 1994 and 2004, and +1.81% between 2004 and 2014. During the combined periods, from 1994 to 2014, Boufekrane and Ouislane experienced significantly stronger than average growth, tripling their population. Boufekrane is a small city, 15 km south of the centre of Meknes, while Ouislane is an eastern suburb that increased its percentage of the region's population from 6% to 12%, absorbing a third of the region's demographic growth during the last decade.

A study of spatial evolution in the types of land use based on SPOT satellite images (Valette et al. 2013), allowed us to identify the extent of urban surface area in 2001 and 2011. A comparison of surface area illustrates the increase in built surface area as metropolitan Meknes absorbed more space, farmland in particular, to accommodate a growing population. The urban areas correspond respectively to an estimated 566,000 inhabitants in 2001 and 679,996 inhabitants in 2011. This

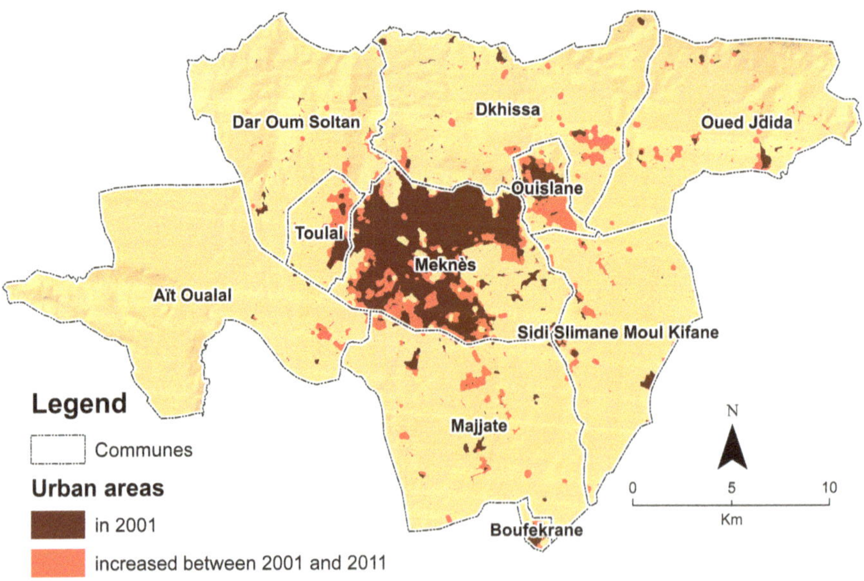

Fig. 4.6 Meknes and its surrounding region, the evolution of urbanization from 2001 to 2011

comparison shows a significant reduction in the density of population on developed land during the studied period, dropping from 183 to 166 inhabitants per developed hectare. Thus the urban spread was accompanied by a concurrent increase in population and decrease in density.

This spread occurred primarily on the agricultural plain of Saïs found on the eastern, southern, and western perimeters of urban Meknes (Fig. 4.7).

The city of Fès lies at the north-eastern limit of the Plain of Saïs; in 2010 it had a population of more than one million inhabitants. Historically, this fertile plain was exploited in pasture or non-irrigated crops and held collectively (guich) in tribal military jurisdictions up until colonial powers took control at the beginning of the twentieth century (Récalt et al. 2013). During the colonial period, the plain was rapidly monopolized by French farmers (Noin 2008), who controlled large farms of cereal crops, orchards, vineyards, and livestock. These colonial lands were nationalized after Moroccan independence in 1956, and partly redistributed as cooperatives by subsequent agrarian reforms. The Meknes region still has approximately 20 cooperatives held as collective, public land in the periurban space of several communities.

The relationships between the urban area of Meknes and urban and periurban agriculture are shaped by two distinct levels of logic: the national level of the Moroccan State, which defines general policies and the terms of their application; and the local level, which interprets public policies in the context of the Meknes metropolitan area.

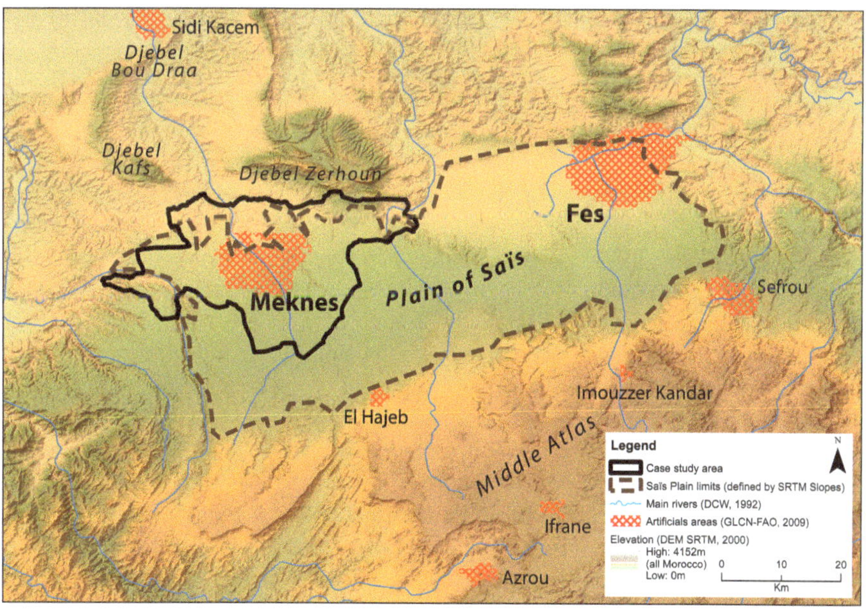

Fig. 4.7 Meknes case study and its geographical environment, Fes and the Plain of Saïs

4.3.2 Morocco and Public Policies by Sector: The Separation of Urban and Agricultural

National public policies have obvious local impacts on the relationships between cities and agriculture. The Green Morocco Plan adopted in 2008 is designed to favour agricultural development in an effort to reinforce production for export and for Moroccan food independence. An apparent preoccupation for food security beyond simply an increase in production for exportation seems to involve the tenuous links between local agriculture and the local food market for consumers.

Moroccan initiatives for urban improvement, such as "Cities without Slums" (Villes Sans Bidonville – VSB) which was implemented in 2004, are primarily aimed at reducing illegal and substandard residential construction to provide new social housing for marginalized populations. Public policies like this also provide support for intermediate housing projects that offer major opportunities for national developers (François et al. 2013). They clearly affect land use and land values in peripheral zones, pushing agriculture further from the urban centres.

The perversion of objectives in two national mechanisms appears to play a fundamental role in the expansion of urban development on farmland:

- The privatisation of collective farmlands through the agrarian reform of cooperatives. This process is justified by the hope that individual property rights will stimulate farmer's investment in the land. (Daoudi 2011).
- The mechanism for exemptions, case by case in exceptional circumstances, for the approval of the use of farmland for urban development (housing, industry and commerce), deviates from the local regulations of land use.

Thus, in spite of regulations aimed at preserving farmland near cities (such as law 12–90 which provides guidelines for urban planning), urbanization has not been significantly restricted (François et al. 2013).

As a result of the strong sectoral aspect of public action (Valette et Philifert 2014), the urban and agricultural systems are usually considered separately. We can use a causal diagram to represent the anticipated benefits of various sectoral public policies implemented in an effort to globally reinforce Morocco's economic and social development (Fig. 4.8).

In the urban sector, development supported by policies to improve housing has a positive impact on the lives of the inhabitants and increases employment. Figure 4.8 shows the circles of anticipated benefits formed by two positive feedback loops. The resulting social and economical development reinforces the legitimacy of these public policies, maintains their positive image and ensures their perpetuation in urban areas. The anticipated material benefits to living conditions are presumed to produce and support general improvements in social and economic conditions: the housing is new, there is little risk for the banking system, and residents can invest in various forms of collective or individual housing developments where ownership is

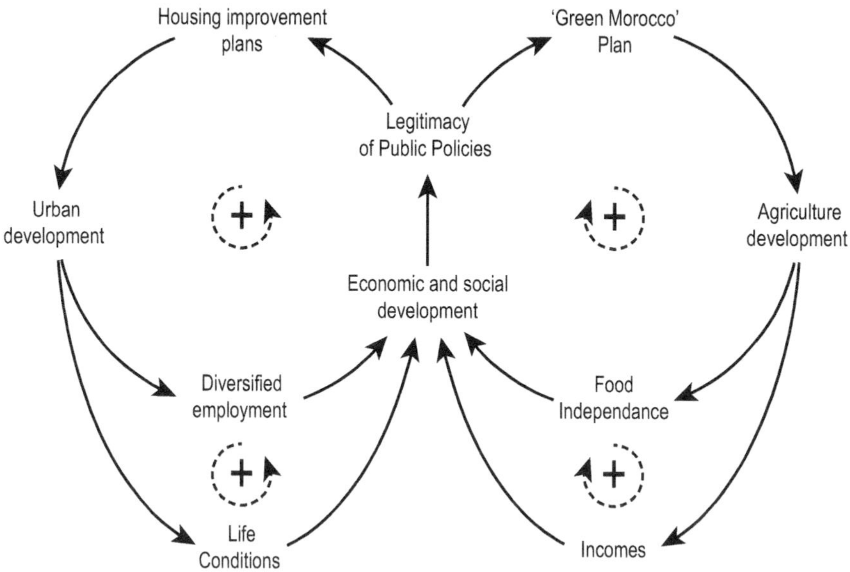

Fig. 4.8 The positive circles of anticipated benefits from Moroccan public development policies at the national level

official and secure. In the agricultural sector, development through the Green Morocco Plan anticipates an increase in revenues for farmers as a result of greater yields and better distribution of products. The plan also seeks to reduce the risks of famine by reinforcing independent food security at the same time as it contributes to economic and social development.

However, the lack of direct relations between urban and agricultural development, as seen in Fig. 4.8, questions the existence of both an explicit will to preserve farmland (at least in official documents), and the recognition of the various social, recreational, and food production functions provided by urban and periurban agriculture. At best, Moroccan public policies view undeveloped urban space through a perspective favouring measures such as natural parks and periurban forests, but this view remains primarily shaped by the larger framework focusing on the environmental issues of sustainable development (Valette et Philifert 2014). The system of interactions between cities and agriculture has not been addressed by national public policies. But shouldn't this system be considered at the local level where urban and agricultural development compete for access to land and relate directly to the regulatory processes?

4.3.3 Meknes Land-Use Dynamics: The Elements of a System of Local Interactions Between the City and Agriculture

In 2001 Meknes authorities established an urban planning policy, the Schéma Directeur d'Aménagement et d'Urbanisme (SDAU). The policy (currently being revised) promised to preserve a green belt at the south-western perimeter of the city with the specific objective of protecting urban and periurban agriculture on land recognized as particularly fertile. In principle, this policy fixed a limit to the urban expansion of the city, but in fact development has not followed the recommendations of the SDAU (François et al. 2013).

The absorption of agricultural land by urbanization remains a phenomenon generally supported in Meknes and throughout Morocco. Satellite imagery reveals that urbanization in Meknes and along its perimeter consumed more than 1000 ha between 2001 and 2011, increasing the urbanized area from 3083 ha to 4087 ha (Valette et al. 2013).

Over the last 15 years the Moroccan State government has established a vast land reserve in the form of New Urbanization Zones (Zones d'Urbanisation Nouvelles – ZUN), destined to encourage local involvement in development policies that support new housing. This program has created 12 of these urbanization zones on state-owned land in the area of Meknes, representing 1303 ha. Additionally, agrarian reforms that transferred cooperative land to private ownership have enabled a significant number of farmers to profit from selling farmland to private urban developers (François et al. 2013).

Another element, the 1999 mechanism providing for exemptions, has made building permits simpler and easier to obtain. Over the span of a decade this mechanism facilitated 338 projects in Meknes. It provides a means of circumventing Meknes' SDAU and reinforces the dynamic of urban expansion onto available land, which is creating new areas of development on the outskirts of Meknes, particularly in the communities of Toulal, Ouislane, and Majjate (Figs. 4.9 and 4.10).

The augmentation in housing follows the intentions of public policies designed to address the social issue of living conditions, but it has also resulted in extensive urban spread. This evolution has been compounded by the strategies of cooperative farmers seeking to profit from the privatisation of agrarian reform (Valette et al. 2013). Although the urbanization of farmland is perceived as a direct menace to the future of agriculture, at the same time it provides landowning farmers with the temptation of a substantial source of income. In addition, the numerous exemptions offered for housing development projects make the sale of farmland even easier. These strategies and their impacts on the evolution of periurban agriculture and land-use in Meknes can be seen in the case of the cooperative Naiji, in Majjate (Valette et al. 2013), where 19 out of 26 farmers decided to sell land they received in a transfer to private ownership in 2012. Almost half of the farmers subsequently abandoned farming activity in favour of commerce, real estate development, or retirement. The rest of the farmers in this group, slightly more than half, invested in farmland further from city, either as a sole activity or in combination with others.

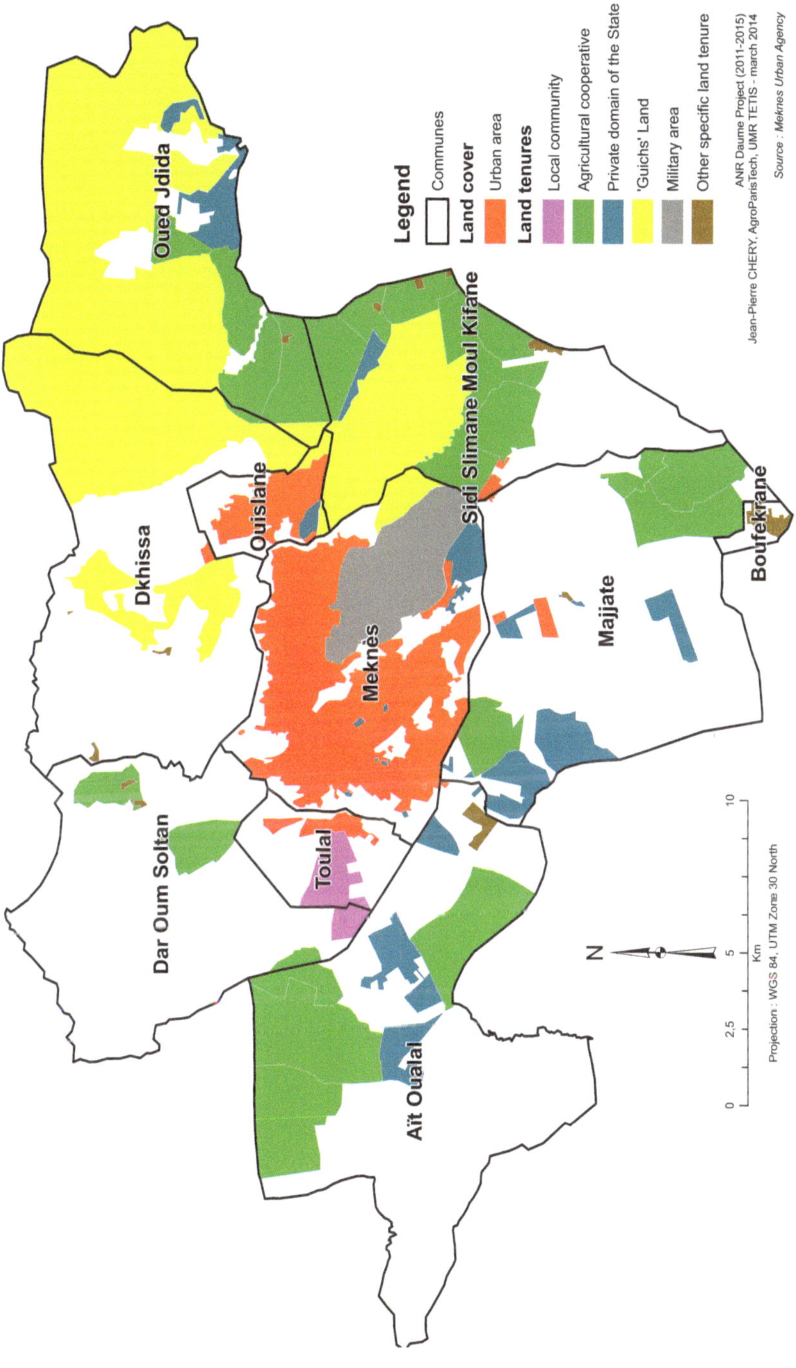

Fig. 4.9 Land-use in the Meknes region at the beginning of the twenty-first century

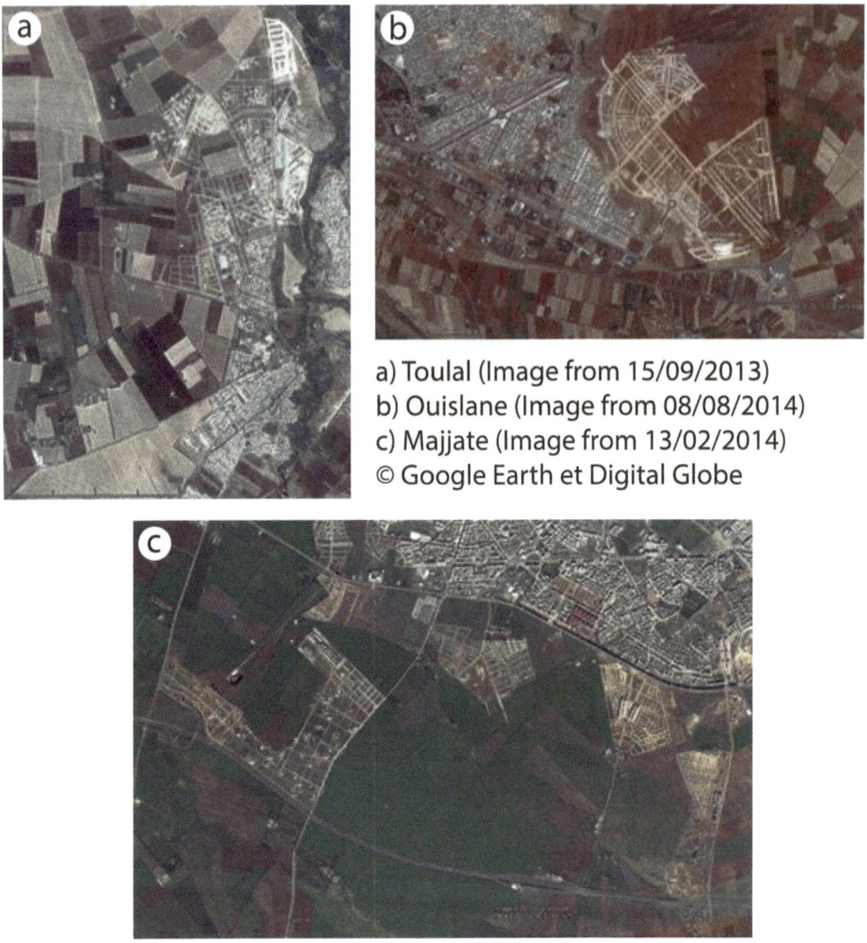

Fig. 4.10 Expansion of urban construction (2013–2014) and associated roadways

Among the remaining seven farmers, two of them stopped farming even though they didn't sell the land, and the others were limiting their activities to crops with low production costs while waiting to sell at a better price. The decisions of these farmers illustrate that the reforms for land privatisation, and the exemptions from policies to contain the region's development have contributed to a dynamic that jeopardizes agricultural activities or displaces them further from urban centers. From this perspective, the apparent priority afforded urban development issues over agricultural issues in the Meknes region suggests that substantial challenges remain for any goal of sustainability in the system of local agricultural and its relationships with the city.

4.4 Interactions Between Meknes and Agriculture: An Unsuspected Number of Relationships

We developed a causal graph to establish the relationships between different elements seen as significant in the system of interactions between the city of Meknes and local agriculture. These relationships were subject to the constraints of national public policies and regulations, and correspond to the processes active during the period from 2010 to 2015 (Fig. 4.11). The causal relationships between 34 chosen elements reveal 13 feedback loups, six of which were amplifiers and seven that were regulators.

Beyond the phenomena (pictured in Fig. 4.11) that diminish agriculture's presence in the Meknes urban region, we analysed two additional loops (bold, black and grey arrows in Fig. 4.12).

4.4.1 The Liberalization of Farmland, or Privatization and Environmental Externalities

An initial amplifying loop highlighted in Fig. 4.12 (bold black lines) illustrates the different fields of sustainable development. This loop can be approached from any of nine elements that follow one another: priorities of public policy, liberalization of agricultural land-use/ownership, investments, modernized farming, intensive farming, pressures on the water table, health issues, and social peace. We have observed that the exploitation of water resources can be seen as one economic causal factor linked to land and agriculture. The phenomenon of seasonal water shortage, particularly in the dry riverbeds of urban Meknes, forces market farmers to use the network of waste water to maintain crops. This practice, along with the frequent use of pesticides on the agricultural plane, creates sanitary problems in agricultural products that undermine the local population's confidence in locally produced foodstuffs (Rachik 2010). This amplifying loop suggests that intensified local agriculture amplifies health issues linked to the water resources, and thus represents one of the weaknesses that undermine sustainability in the system (McMichael 2006).

4.4.2 The Construction of Housing, or One Question Answered

The second and simpler feedback loop represents a balancing structure in the system's dynamics (bold grey lines in Fig. 4.12). This structure is formed by four elements: the need for social housing, exemptions from development restrictions, rapid and uncontrolled artificialization, and urbanization on the plain associated with the reduction of makeshift housing.

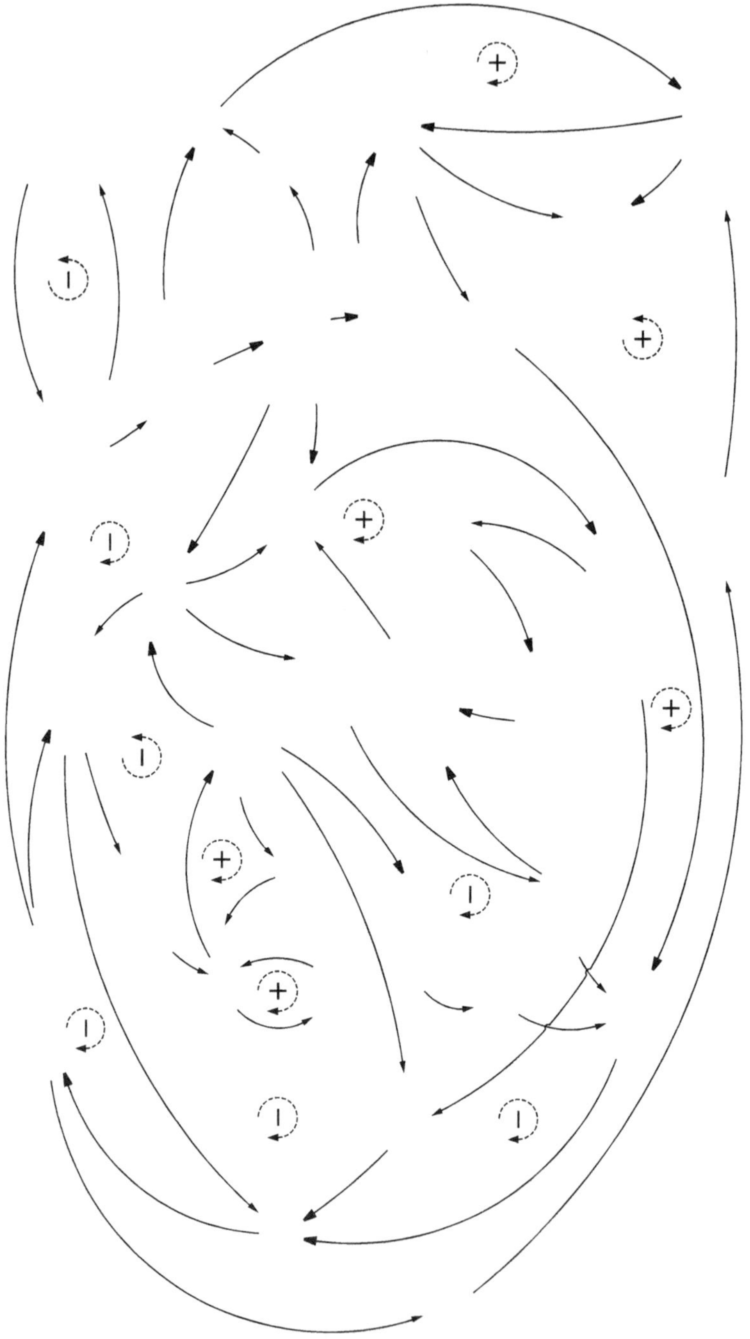

Fig. 4.11 the system of interactions between the city of Meknes and local agriculture

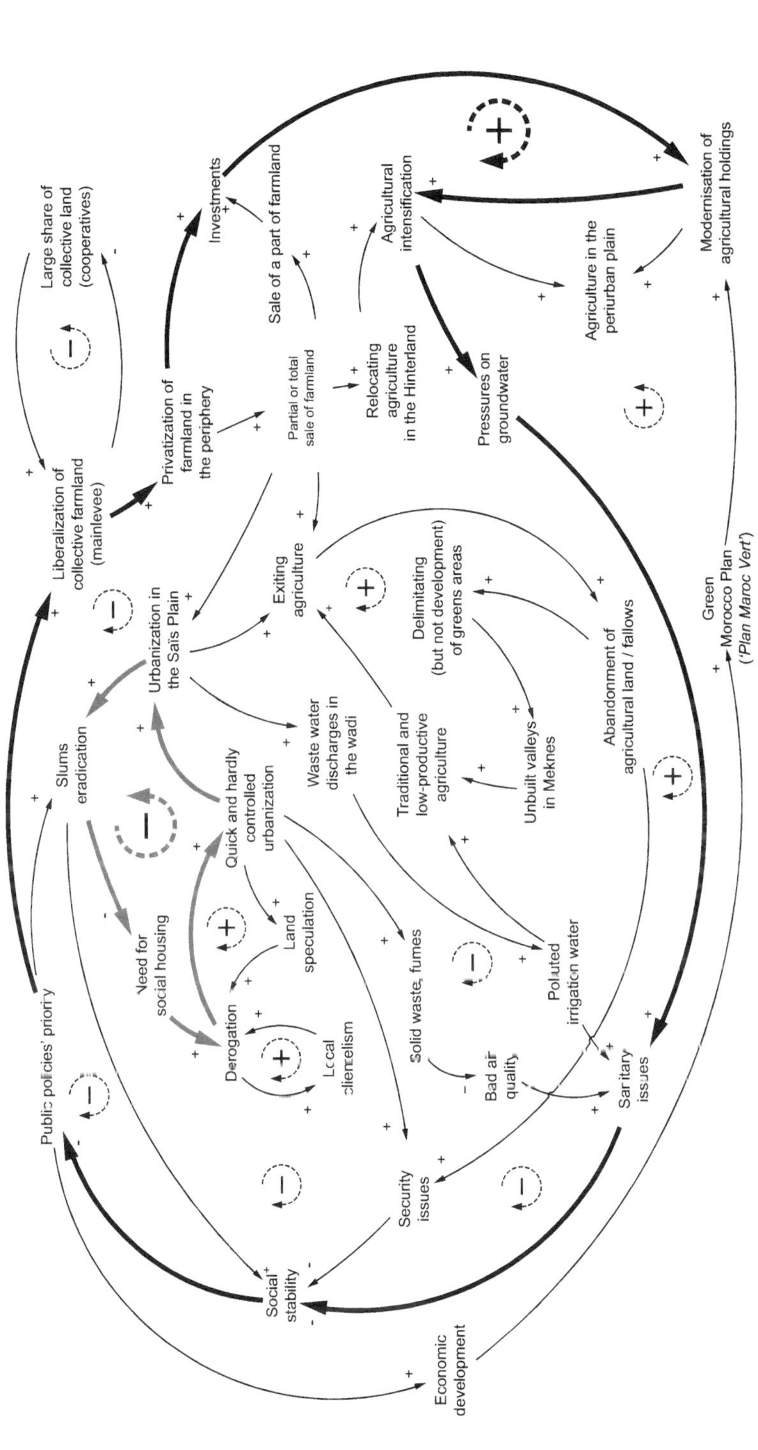

Fig. 4.12 Examples of reinforcing and balancing feedback loops of a complex system of interactions between the city and agriculture

The need for social housing creates a favourable climate for housing development projects that take advantage of the system of exemptions. There is also a significant, un-official mechanism of political favours. The multiplication of advantages results in a virtually unregulated and uncoordinated mobilization of land resources, which encroaches primarily on the agricultural plain. This urbanization with social housing also has the support of policies for the reduction of makeshift housing. In this context, we can assume that construction has met or even exceeded actual housing demand. In this case, the demand is met at the expense of farmland, and underlines questions about the sustainability of the system of relationships between the city and local agriculture, and its place within the logics of national public policy.

4.5 Conclusion

The systemic approach allows us to identify externalities that affect both public policy and local action. Feedback loops help describe causal chains that reintegrate elements that have been ill perceived, disregarded or avoided, such as the environmental concerns that appear to be lacking or meagrely addressed in public policies. Much of the language promoting sustainable development has been formulated from the principles of a prophecy that wants to be self-directing without actually insuring its realization. Suggesting the a priori existence of a working system of relationships between city and agriculture in the Meknes region does not prejudge a shared identification with that system among the region's actors. The example of exemptions allows us to underline the lack of basic coordination in governing the regulation: the fragmentation of decisions in the exemption process, amplified by a unofficial system of political favours, reinforces urban spread across agricultural land.

Farmland can be an implementation tool of public policy that operates through various processes. However these processes lead to thresholds beyond which sustainability can be threatened, such as the reserve of public land, which is not inexhaustible, as well as land with potentially strong agronomic value.

The system of interactions between city and agriculture in the Meknes region as it functioned in the first decade of this century does not appear to be sustainable. The on-going sectoral approach observed in the economic and land-use decisions, as well as a lack of coordination, compromises the sustainability of the system. This underlines the need for an 'ecology of action' (Hubert 2004), in order to bring the defined and applied actions to a more global and national scale that addresses the environment of constraints and counter-intuitive relationships.

It seems important to question the clear un-sustainability rather than the potential or desired sustainability espoused by public policies. Effectively defining the actions that can regulate the system of interaction between a city and agriculture appears to require the consideration of the system both on the global scale of national policies, and on the local scale.

Acknowledgement This research has been funded by the French National Agency for Research. DAUME project under n° ANR-2010-STRA-007-01.

References

Akesbi N (2012) Une nouvelle stratégie pour l'agriculture marocaine: le "Plan Maroc Vert". New Medit 11:12–24

Ambrosio-Albalá M, Bastiaensen J (2010) The new territorial paradigm of rural development: theoretical foundations from systems and institutional theories. Institute of Development Policy and Management, University of Antwerp

Berazneva J, Lee DR (2013) Explaining the African food riots of 2007–2008: an empirical analysis. Food Policy 39(0):28–39

Daoudi A (2011) La régulation foncière au Maroc. In: Elloumi M, Jouve A-M, Napoléone C, Paoli J-C (eds) Régulation foncière et protection des terres agricoles en Méditerranée. Options méditerranéennes: Série B., 66, Centre international de hautes études agronomiques méditerranéennes, cop, Paris, pp 63–72

Debolini M, Valette E, François M, Chéry JP (2015) Mapping land use competition in the rural–urban fringe and future perspectives on land policies: a case study of Meknès (Morocco). Land Use Policy 47(0):373–381

Dugue P, Benabed A, Abdellaoui EH, Valette E (2015) L'agriculture urbaine à Meknès (Maroc) à la croisée des chemins: disparition d'une agriculture marginalisée ou retour de la cité jardin? Alternatives Rurales, 3. http://alternatives-rurales.org/809/ [consulté le 05/09/2015]

Durand, D (2006) La systémique, 10e éd. Presses Universitaires de France, Paris

FAOSTAT (2015) Food and agriculture organization of the United Nations. Statistics Division. http://faostat3.fao.org [consulté le 03/04/2015]

François M, Valette E, El Hassane A, Debolini M (2013) Urbanisation des terres agricoles: ressorts, dynamiques, et impact sur l'agriculture à la périphérie de Meknès. Maghreb-Machrek 215:123–140

Gazel H, Harre D, Moriconi-Ebrard F (2011) L'urbanisation des pays du Moyen Orient et de l'Afrique du Nord (MENA), 1950 2030. Tableau de Bord Maroc, Etude MENAPOLIS e-Geopolis, mai 2011. [consulté en ligne: http://e-geopolis.eu/menapolis/TABLEAU_BORD_MAROC.pdf]

Haut-Commissariat au Plan (2015) Population légale des régions, provinces, préfectures, municipalités, arrondissements et communes du royaume d'après les résultats du RGPH 2014, Recensement Général de la Population et de l'Habitat (RGPH), Royaume du Maroc. [Consulté en ligne: http://www.hcp.ma/downloads/RGPH-2014_t17441.html]

Hodge RAT (1996) A systemic approach to assessing progress toward sustainability. In: Robinson J, Dale A (eds) Achieving sustainable development. University of British Columbia Press, Vancouver, pp 267–292

Hove M, Ngwerume ET, Muchemwa C (2013) The urban crisis in sub-Saharan Africa: a threat to human security and sustainable development. Stab Int J Secur Dev 2(1). https://doi.org/10.5334/sta.ap

Hubert B (ed) (2004) Pour une écologie de l'action. Savoir agir, apprendre, connaître. Editions Arguments, Paris

INAU (2005) Évaluation de l'utilisation des terres agricoles à des fins non agricoles. Ministère de l'Agriculture, du Développement rural et des Pêches maritimes, Rabat

Mbiba B (2000) Urban agriculture in Harare: Between suspicion and repression. In: Bakker N et al (eds) Growing cities, growing food. Urban agriculture on the policy agenda. Deutsche Stiftung für internationale Entwicklung, Feldafing, pp 285–301

McMichael AJ (2006) Population health as the 'bottom line' of sustainability: a contemporary challenge for public health researchers. Eur J Public Health 16(6):579–581

Meadows DH et al (1972) The limits to growth. New York,102, 27

Ministère de l'Agriculture (2008) Plan Maroc Vert: Premières Perspectives sur la Stratégie Agricole. Ministère de l'Agriculture, Rabat

Noin D (2008) Saïs. Encyclopædia Universalis [consulté sur: www.universalis.fr/encyclopedie/sais/]

Obosu-Mensah K (2002) Changes in official attitudes towards urban agriculture in Accra. Afr Stud Q 6(3):19–32 [Online] URL: http://web.africa.ufl.edu/asq/v6/v6i3a2.htm

Pearson LJ et al (2010) Sustainable urban agriculture: stocktake and opportunities. Int J Agric Sustain 8(1–2):7–19

Philifert P (2011) La construction d'une politique de développement durable au Maroc: principes, traductions et contradictions. In: Barthel P-A, Zaki L (dir.) Expérimenter la « ville durable » au sud de la Méditerranée. Chercheurs et professionnels en dialogue. Editions de l'Aube, pp 97–120

Rachik A (2010) Nouveaux mouvements sociaux et protestations au Maroc, IRES (Institut Royal d'Etudes Stratégiques), 65 pages

Récalt C, Rouvière L, Mahdane M, Errahj M, Ruf T (2013) Aménager l'espace, canaliser l'eau et orienter le pouvoir. Réflexion sur deux modèles inédits d'aménagements fonciers radioconcentriques en France et au Maroc. Autrepart 65(2):107–127. https://doi.org/10.3917/autr.065.0107

Valette E, Philifert P (2014) L'agriculture urbaine: un impensé des politiques publiques marocaines ? Géocarrefour 89(1–2):75–83

Valette E, Perrin C, Soulard C (2012) Sustainable cities vs sustainable agricultures. A scientific project on agro-urban systems, North and South of the Mediterranean. In: International conference on multifunctional agriculture and urban-rural relations "agriculture in an urbanizing society". Wageningen University, The Netherlands, 1–4 April

Valette E, Chéry JP, Azodjilande J, Debolini M, François M, El Amrani M (2013) Urbanisation en périphérie de Meknès (Maroc) et devenir des terres agricoles: l'exemple de la coopérative agraire Naïji. Cahiers Agric 22:535–543

Voiron-Canicio C (2005) Pour une approche systémique du développement durable. Développement durable et territoires [online], Dossier 4-2005. URL: http://developpementdurable.revues.org/261; https://doi.org/10.4000/developpementdurable.261

World Commission on Environment and Development (WCED) (1987) Our common future. Oxford University Press, Oxford\New York

Yemmafouo A (2014) L'agriculture urbaine camerounaise. Au-delà des procès, un modèle socioculturel à intégrer dans l'aménagement urbain. Géocarrefour 89(1):85–93

Chapter 5
Urban Farms Under Pressure: Cairo's Dairy Producers, Egypt

Annabelle Daburon, Véronique Alary, Ahmed Ali, Mohamed El Sorougy, and Jean François Tourrand

Abstract The rapid increase in population of Egypt's Nile Valley and Delta is leading to massive urbanisation and uncontrolled conversion of agricultural land. Food security appears to be a daily challenge in cities all over the country. Greater Cairo is home to approximately 20 million inhabitants. Most of the city's expansion is occurring on arable land, absorbing outlying farm systems and questioning their sustainability. In an exploratory study of dairy farming systems in urban and peri-urban areas of the capital, four main types have been identified by their capacity to integrate crop and livestock production, and by their structural characteristics. These farming systems are facing several primary constraints: limited space and access to fodder, increases in feed concentrate prices, and manure management. Despite this difficult context, they are maintaining their activity by serving the community and taking advantage of its proximity: They are incorporating adaptive strategies for feeding systems, breeding systems, and inputs sourcing, and providing employment, food security, and waste recycling. These strategies appear to be initiated mainly at the scale of family farming systems. Although these systems are

A. Daburon (✉)
CIRAD, UMR SELMET, SELMET, University of Montpellier, CIRAD, INRA, SUPAGRO, Montpellier, France
e-mail: daburonanna@gmail.com

V. Alary
CIRAD, UMR SELMET, ICARDA, Rabat, Morocco

SELMET, Univ Montpellier, CIRAD, INRA, SUPAGRO, Montpellier, France
e-mail: veronique.alary@cirad.fr

A. Ali · M. El Sorougy
Animal Production Research Institute, Cairo, Egypt
e-mail: alitamatem@yahoo.com; m_a_elsrogi@hotmail.com

J. F. Tourrand
CIRAD, UPR GREEN, Montpellier, France

GREEN, Univ Montpellier, CIRAD, Montpellier, France
e-mail: jean-francois.tourrand@cirad.fr

C.-T. Soulard et al. (eds.), *Toward Sustainable Relations Between Agriculture and the City*, Urban Agriculture, https://doi.org/10.1007/978-3-319-71037-2_5

evolving rapidly, their future survival remains questionable in a context of the evolving demands of one of the world's largest megalopolises.

Keywords Family farming · Urban agriculture · Adaptability · Milk production

5.1 Introduction

Egypt is a southern Mediterranean country with a rapidly growing population of more than 93 million inhabitants. This population is concentrated in the same areas as the country's stock of arable land, lying primarily along the Nile River (El-Nahrawy 2011). One of the largest African city emerged during the last century: the megalopolis of Greater Cairo with 20 million residents. Some of Greater Cairo's informal settlements have reached a population density of 150,000 people/km^2 (Séjourné 2006). The expansion of cities across arable land is an on-going and seemingly uncontrolled process throughout the country generating conflicts within the population (Johannsen et al. 2009). Although the conversion of arable land is restricted by law, the state has been unable to fully protect its arable land resources (4% of the country's surface area, or 4 million ha). In an effort to counter this trend, public policies such as the New Reclaimed Land (NRL) strategy have been implemented to transform desert land into agricultural areas, mainly on the border of the Delta (El-Nahrawy 2011; Adriansen 2009).

According to the World Bank, Agriculture sector constituted almost 15% of the gross domestic product in 2014. Small family farming systems realise the majority of the national production of cereals, vegetables, and animal products. They are generally irrigated systems with limited structural resources that integrate both crops and livestock production. The dairy sector is no exception; national milk production is provided mainly by family farms with small herds (a majority having no more than 3 or 4 dairy animals), mixing cow and buffalos. An industrial sector in Egypt's dairy industry has been gradually emerging over the last 30 years. National and international companies, attracted by increasing demand and favourable policies (regarding the land access and financial markets) in the NRL have developed industrial farms with large herds (thousands heads) on those territories (Dixon 2014). Egypt's fresh milk supply is complemented by imported milk powder (1.6 million tons of milk equivalents were imported in 2012, mostly milk powder) (FAO) that furnishes raw material for standardized industrial dairy products.

A large segment of Egypt's population is gradually adopting an urban lifestyle. This change includes an evolution in diet, with an increase in the total calories ingested per capita and a shift towards a larger proportion of animal compounds in the diet, both in meat and milk consumption (Popkin and Wen Ng 2007). The general population increase, coupled with the evolution in demographics and diet, has resulted in a rapidly growing demand for animal products in the food supply. Facing an annual consumption of dairy products (excluding butter) of over 64 kg per year per capita in 2011 (Alary et al. 2016), the milk supply of Greater Cairo has become a daily challenge. According to sources (economic experts, private companies, and the ministry of agriculture), between 65% and 85% of the raw milk market was supplied by "loose milk" in 2012–2013. This term is used in Egypt to characterize liquid milk produced by family farming systems and distributed through a dense and complex variety of intermediaries and marketing pathways. The majority of the milk is produced throughout the Delta, but some of these family farming systems are located directly in the periphery and even well inside cities, including the Greater Cairo megalopolis. The lack of data regarding urban and peri-urban agriculture (UPA) in this Egypt, as in many cities (Padgham et al. 2015) means that authorities show little consideration or interest concerning those producers. At this point we have been unable to properly identify a law that forbids livestock inside cities and more specially inside Cairo's urban space, but with more than 60% to 80% of the city's surface considered as informal areas (unplanned and uncontrolled) that have gradually expanded over agrarian land (Kipper and Fischer 2009; El-Hefnawi 2005), the regulations appear to be local and informal.

UPA is an emerging concern directly linked to the sustainability of cities, food security and a potential for climate change mitigation (Padgham et al. 2015). Several studies have been conducted highlighting the role of UPA in major cities (Sabiiti et al. 2014; Gyasi et al. 2014; Robineau 2013). Although this role can be challenged (Warren et al. 2015), characteristics and adaptation mechanisms that contribute to the sustainability of UPA in a highly constraining environment, remain a rich field of investigation. The case of family farms in Cairo's UPA presents an opportunity to study these adaptation mechanisms and their capacity to help sustain a system that contributes significantly to the food supply in one of the world's largest megalopolises. How do sustainable UPA systems adapt and optimize their interactions with the megalopolis environment? This is the main question that this chapter will explore. We will begin with a description of the family farming systems, and then consider the links that exist between the city and its farmers, and how the integration of farming within Cairo has evolved. We will then consider the mechanisms involved in the adaptive ability of Cairo's UPA, and highlight the relevant perspectives and challenges.

5.2 Materials and Methods

A cluster analysis of the dairy farming system in urban and peri-urban areas of Greater Cairo was performed to identify the diversity of situations and practices adopted. During 3 months (March to June 2013), researchers used a semi-structured questionnaire[1] to conduct 73 interviews of farmers as a part of the Dairy project:

[1]Topics investigated: (i) Social: history, family composition, level of education of the head of household; (ii) Cropping system: total cultivated area, land access, crop rotation system, inputs costs; (iii) Livestock system: herd description, total Tropical Livestock Unit (TLU: 1TLU = 1 cow

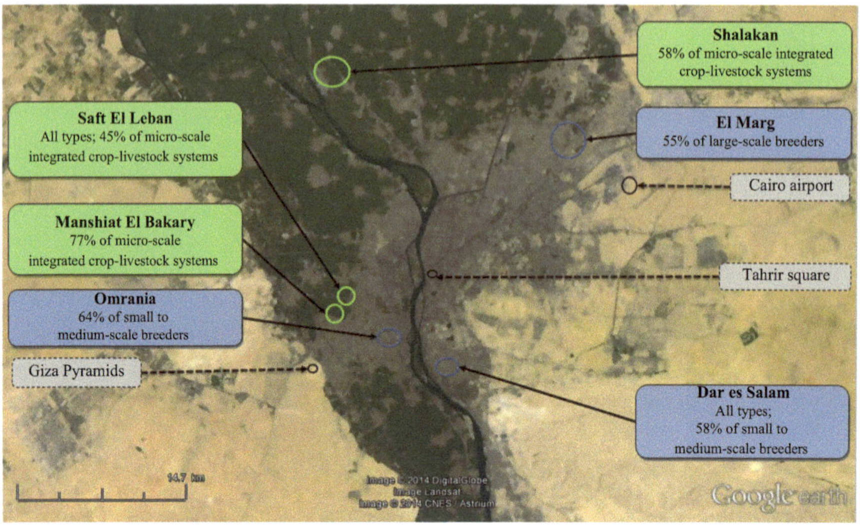

Fig. 5.1 Cairo satellite picture and research areas. In blue, urban area; in green, peri-urban areas

Understanding the traditional milk supply chain of Greater Cairo (AIRD Young team, 2012–2015).

Several areas in Greater Cairo appear to contribute to the milk supply and six areas were selected according to our opportunities to secure contact with at least one producer: three were chosen well inside the city and three in outlaying areas of the city (Fig. 5.1).[2] The only exclusive condition in the selection of farmers was the possession of one productive dairy animal during the year prior to the visit. A "snowball sampling" technique (Goodman 1961) was applied within each zone, due to the lack of farm registration. This technique is based on the social network of the subject guiding the researcher to his next subject, until 12 farmers are interviewed per area (13 in one). Based on the data, a set of variables enabled a factorial analysis (Hill and Smith 1976), followed by a cluster analysis (Anderberg 1973) through a Ward clustering method using open-source software R, package ade4 (Dray et Dufour 2007). These steps enable the construction of a typology of the UPA systems (Landais 1996). These tools are used in a system approach that considers the farm as a complex system composed of sub-entities interacting with each other and with the totality in a holistic way (Brossier 1987). Due to the methodology and the lack of prior data, these results need to be considered as exploratory research that identifies trends, general organisations, adaptation mechanisms and the primary relationships binding UPA systems with their urban environment.

with a body weight of 250 kg), demography of the herd, feeding system, milk production, product marketing; iv) Problems and perspectives.

[2] Qualification of urban or peri-urban areas, in this sample is mainly indicative. It was set according to the localization of the family house: included inside the city for urban or close to the city limits for peri-urban.

5.3 Results

5.3.1 The Diversity of Urban and Peri-Urban Farming Systems

All of these farms belong to the model of a family farming system as described in Daburon et al. (2014). The first characteristic that differentiates these family farms is whether or not they integrate crop and livestock production. Although some families are cultivating land, others have no access to arable land, and rely only on their indoor animal production. The next criteria defining farms are their structural characteristics (main descriptive characteristics are presented in Table 5.1).

In all of these systems, production (crops and animal products) first serves to fulfil family and herd requirements. Marketing only takes place if a surplus is produced.

Table 5.1 Main variables describing farming systems in Cairo urban and peri-urban areas

	Small-scale integrated crop-livestock system	Micro-scale integrated crop-livestock system	Large-scale breeders	Small to medium-scale breeders
Percentage of the sample and number of farms	21% (15)	33% (23)	17% (12)	29% (20)
Main purpose of farm activity	Market	Self-consumption	Market	Market
Number of adult dairy buffalo in the herd (average)	5,54 (σ = *4,8*)	1,3 (σ = *1,1*)	26,4 (σ = *31*)	6 (σ = *4*)
Number of TLU[a] per farm (average)	21,5 (σ = *15,9*)	9,7 (σ = *6,7*)	73 (σ = *64*)	17,9 (σ = *11,6*)
Cultivated land area (rented or in property) feddan[b](f.)	3 (σ = *3,5*)	1 (σ = *2,3*)	0 (σ = *0.6*)	0 (σ = *0.4*)
Feed cost for a dairy animal/day (€)	2,6 (σ = *0,8*)	1,8 (σ = *0,7*)	4 (σ = *0,7*)	3,3 (σ = *0,9*)
Median gross margin[c] from agricultural activity per family member(including home consumption)(€)	105 (σ = *203*)	24 (σ = *47*)	393 (σ = *225*)	*−5,3* (σ = *0,6*)
Number of family members per farm	13 (σ = *13*)	14 (σ = *9*)	10 (σ = *11*)	8 (σ = *8*)

[a]TLU: Tropical Livestock Unit, several coefficients: Sheep 0,35; goat 0,23; donkey 0,85; camel: 1,68; horse 1,68; buffalo female 1,8; baladi female 1,7; bull 2,1; pregnant heifer 1,6; calf 0,4; fattening calf 1,3. (FAO) http://www.fao.org/ag/againfo/programmes/en/lead/toolbox/Mixed1/TLU.htm

[b]Feddan: surface unit in Egypt, 1 feddan (f.) = 0.42 hectare (ha.) = 1.038 acres (ac.)

[c]Gross margin from the agricultural activity (crop and livestock productions), including opportunity costs of home-consumption

5.3.1.1 Integrated Crop-Livestock Systems

Common Characteristics of the Farms These family farming systems rely both on animal and crop production. Agricultural plots may be owned, as part of the farm property, or rented, or be a mixture of the two types of land access. Several crops are generally planted in a similar pattern among the farms:

- Fodders covering 50% of the land: clover (*Trifolium alexandrinum*) in the winter, and maize fodder (*Zea mays*) in summer. This production was used to feed the herd (surplus can be sold).
- Cereals covering 30% of the cultivated area: planted with wheat (*Triticum aestivum)* in winter, and corn (*Zea mays*), or in rare circumstances, rice (*Oryza sativa)* in summer. Grains may be kept for family and animal consumption or sold to intermediaries. Straws (wheat and maize) are kept for feed and litter or sold. Rice straw is often burned in the field, but new practices are appearing (use as feed, raw material for paper production).
- Cash crops (mainly vegetables) on the remaining 20% were planted with vegetables. A large variety of vegetables is produced throughout the year: tomatoes, onions, cucumbers, aubergines, etc.

Dairy buffalos constitute the bulk of their animal production. Fattening activities could be identified, but remained complementary in the farms of our sample. A lack of public services or cooperatives means that most of the feed concentrates are bought from outside sources, mainly private dealers. Animals spend the night on the ground floor of the family building or in a closed, dedicated building nearby (justified by farmers as security against theft due to the high value of animals). After morning milking and distribution of feed, a family member may move the dairy animals to the family agricultural plots if access is easy (less than 4 km with limited urbanization). They spend the day, attached, during field work, in a small wood shelter, receiving fodder cut by a family member twice a day. In the evening they are returned to the family building for the second milking and concentrate distribution. When the animal stays at home, fodder supplied from a distant field, constitutes an important time-consuming task of family labour.

Small-Scale Integrated Crop-Livestock Systems This group represents roughly 21% of the total sample. They cultivate larger plots of land, generally close to 3 feddan (1.26 ha.), of which an average of 40% is owned by the farm. The balance of land is rented (see Sect. 3.2.1). In animal production, these families hold larger herds of almost 21.5 TLU, including 5–6 adult dairy buffalos along with a few sheep or goats and sometimes one or two baladi[3] cows. Daily feed cost for a dairy animal was estimated at €2.6 per day per head[4] and almost all of the fodder is produced on their land. With more than 13 members per family, this group is second in family

[3] Baladi cows: a local breed, raised for meat and milk, but often mixed with other genetic (primarily Holstein) in varying proportions.

[4] Conversion rate of Egyptian Pound (EGP) to Euro in May 2013: 1Euro = 9.0767 EGP.

size, and their farms generated a gross profit margin of approximately €105 per month per family member, 61% from animal production.

Micro-Scale Integrated Crop-Livestock Systems These producers also integrate crop and livestock production, but with smaller assets than the previous group. They represent 33% of the total sample, and their cultivated area reaches 1 f. (0.42 ha), with approximately 80% on land owned by the farm. In combination with this cropping activity, they keep small herds, 9.7 TLU, with 1 or 2 adult dairy buffalos. They have the lowest feed cost for dairy animals in our sample with only €1.8 per day per head. With 14 members, they had the largest families in the sample, but their gross profit margin from farming only reaches €24 per month per member, including 50% from animal production.

5.3.1.2 Indoor Livestock Farming

These farming systems subsist without a cultivated area and rely only on animal production (see Sect. 3.2.2) (Fig. 5.2). The animals live under houses or buildings, or on the ground floor. Dairy buffalos are the most common species in these

Fig. 5.2 Pictures from indoor livestock breeders: top left, calf on the ground floor of a house; top centre, Cairo buffalo farm; top right, farm entrance (40 adult buffalos and cows behind the door); bottom centre, dairy buffalo from Cairo; bottom right, straw truck between buildings in a Cairo informal area

systems, although some cows or small ruminants may be kept for fattening. These farmers can also be distinguished by their structural characteristics: large breeders and small to medium breeders.

Large-Scale Breeders These families represented 17% of the sample, and own the largest herds of the sample, 73 TLU, with roughly 26 adult dairy buffalos. Their daily feed costs per animal are the highest at €4 per day per head. Despite the high feed costs they have the highest gross profit margin: €393 per month per member, in families with ten members.

Small to Medium-Scale Breeders Finally, the 20 remaining farms, from the sample have limited herds of 17.9 TLU with six adult dairy buffalos. The daily feed cost remains high compared with the rest of the sample: €3.3 per day per head. This group has the smallest families, counting only eight members, and it appears difficult for them to create a profit from their animal activity. This situation may be explained by their inability to negotiate with private dealers and reduce inputs costs, as compared to large breeders that are able to economise in quantities, but this conjecture would require further investigation.

5.3.2 What Are the Main Farming System Adaptations Engendered by the Immediate Proximity of a Megalopolis?

We have identified four types of UPA. Now we will present the principle interactions observed between these UPA systems and examine how farmers have been able to sustain their activity in certain cases by adjusting their system and taking advantage of their proximity to the megalopolis.

5.3.2.1 Threatened by Urbanization, Systems Integrating Crops and Livestock

The decline of soil fertility was probably one of the initial consequences of urbanization. The decrease in efficiency of drain and irrigation networks resulting from the accumulation of garbage, road conversions and road closures, forced farmers to irrigate with saline groundwater, increasing salinization of the soil and in some cases leading to sterile land. In the same time, farmers have been faced with the disappearance of arable land from illegal urban expansion on agricultural land surrounding Cairo. Indeed, during the period of this research, the value of 1 ha of peri-urban land was approximately €500,000. That hectare represented the possibility of developing 12 to 15 buildings of 10–12 floors with around 4–6 apartments per floor, meaning somewhere between 500 and 1000 apartments. Examining these

conditions among families integrating crop and livestock production, two situations can be contrasted depending on ownership of the land.

If the family owned land, then they had the option of taking the profit from selling the land. This capital could then be invested in several ways:

- A plot in the Delta. This case was rare due to the tight market for historically arable land.
- A plot in NRL. Again, opportunities seem to have been rare due to the limited amount of land available and the family was required, at least in part, to move out of Cairo (thus weakening the security of their social network proximity).
- Investing in indoor livestock farming, without land access, which was the main option chosen by large-scale breeders interviewed.
- Abandon all agricultural activity and invest in other businesses.

For tenant families, choices were extremely reduced. If until the end of the 1990's they were protected by law (the rent was limited and tenant prevented to be expelled), nowadays most of the contracts must be renewed every year and the cost of renting land had increased rapidly (average of €1250 per ha. per year, in our sample, around Cairo). In 2013, land owners had the option of selling their plots to promoters, while tenant farmers had no protection, and appeared extremely vulnerable. If a land-owner decided to sell his plots or to increase the rent beyond the affordable limits of the farming families, few options remained for these families to maintain their farming activity:

- Find new plots to rent, in the context of a very tight rental market for arable land. This implies extending the distance between their plot and their houses inside Cairo. Thus increasing the difficulty of maintaining a system integrating crops and livestock, particularly since the majority of these families were using a donkey cart for transportation.
- Abandon agricultural crops, and rely on indoor livestock farming inside the city. Once again, most of the small to medium-scale breeders interviewed in the study correspond to this scenario.
- Their last option was to abandon agricultural activity entirely. But the members of most of these families, have little education or qualification for other work. By abandoning agricultural activity, they would add to the growth of precariousness in the urban areas.

Finally, beside urban expansion, fragmentation of family plots seemed to be a direct consequence of demographical growth affecting these farming systems. With each generation, family plots and herds are divided between heirs, leading to a fragmentation of the capital. In a number of cases, to mitigate this process, the family entrusted the farm system to a single heir. Production or profits were then divided between family members. This phenomenon contributed to improve food security for the extended family and maintain the prestige of land ownership.

5.3.2.2 New Sources of Inputs for Cairo's Livestock Production

With the increasing difficulty of access to fertile land, farmers must import at least a portion of their inputs. The farming systems rely partially or totally on the market for primary inputs such as water, fodder, concentrate and litter. In the case of each of these items they have developed adaptive strategies. These strategies include strong links with the megalopolis that help both fulfil their supply needs, and provide benefits to the city, such as recycling wastes.

The procurement of feed inputs is one of the most adaptive characteristics observed in these farms. This system evolves in a dynamic process depending on feed availability and prices. Cairo's extensive urban expansion has led to a scarcity of fodder which worsens every year. This requires farmers to import their fodder from further distances (sources such as farms on NRL) regularly with lesser quality products. Occasionally some farmers have an agreement with suppliers for lower priced fodder in exchange for their manure (a valuable input for NRL). However, the majority must decrease or even sometime eliminate green fodder from their animals' feed ration. Dry fodder is used in large quantities to balance the scarcity or absence of fresh fodder. Wheat straw is the most common substitute, but new resources have been developed from former waste products. Cairo has a large number of juice producers that press sugar cane (Egyptian products) which results in large quantities of bagasse.[5] In the last few years, enterprising dealers have been collecting bagasse and selling it to Cairo's livestock farmers. Another alternative dry fodder, rice straw, appears to be emerging. Produced in large quantities in Egypt, rice straw has traditionally been treated as a waste product and burned each year after harvest, causing significant atmospheric pollution throughout the country.

Concentrate feeds are generally purchased, even if some of the families interviewed were producing a small portion, predominantly grain maize. The basic components used are wheat bran, grain maize, and industrial concentrate pellets. At the most specialized dairy farms (those with sufficient financial resources) cottonseed and soya beans were introduced. For numerous poor families, adaptive strategies have been developed, such as using stale bread or leftover bakery products, sometimes provided by dealers. Similarly, but less common, waste products from an industrial bakery, or restaurants were used to feed animals.

The availability of water, in quantity and quality, is a crucial point for dairy production. For the majority of these farmers, water for both human and animal consumption is supplied by the diverted public network, more or less illegally. Regular water cuts constituted a major issue forcing some families to dig a well inside their house, questioning the quality of the ground water consumed by animals.

[5] The pulp remaining after the extraction of juice from sugar cane.

5.3.2.3 How to Maintain Sanitary Conditions for Farmers, Their Animals, and Their Neighbours?

The coexistence of animals and humans in a city with one of the highest population densities in the world is a constant challenge. In part because of the increasing difficulty to reach exterior plots, more than 50% of the animals observed were kept indoors and attached at all times, thus raising health concerns.

Herds confined indoors can present numerous health risks. Overcrowded stables with high temperatures and humidity are a favorable environment for diseases if the farmer does not maintain a high level of hygiene management. One of the main challenges for these farmers is the daily removal of manure out of the farm and out of the city. Families with land access, use manure to fertilize the soil, others rely mainly on a private network of dealers that collect manure for free and then sell it (usually to farms in NRL).

An adaptive strategy of reproductive management was also observed. Dairy buffalos and their calves are bought 1 or 2 days after the calving. The young animals, in these Cairo systems, are often sold after weaning (2–3 months) to dealers that supply fattening farms or butchers (veal represents high profit). During a short lactating period (6–7 months), adult buffalos produce an average of 2140 l/year (sample average), however the high ratio of concentrate to fodder carries the risk of creating acidosis. Consequently, the animal is culled after this milking period, and following a very short fattening period, sold to dealers, most of the time for slaughterhouses. This very short turnover within the herd allows the farmers to avoid both the cycle of feeding "unproductive" animals during their dry period, and the risks of keeping animals with impaired health for another lactation. They effectively "burn out" animal resources and health in a final intense milking period, knowing that they will not be engaging in a new reproductive cycle.

The context of unclear sanitary farming conditions also leads to zoonosis issues. Both the farmer's family and the consumers incur a significant risk of zoonotic diseases such as bovine tuberculosis or brucellosis, which is endemic in Egyptian herds. Traditionally, Egyptian consumers boil their milk before consuming it, thus reducing some contamination risks, but breeder's families in constant, direct contact still appear vulnerable. Although a few cases seemed to have difficulty, the majority of farms in the sample maintained a remarkable level of hygiene, considering the context. Steps taken to limit air pollution and noise helped to avoid disturbing neighbours and maintain acceptance by the community while at the same time decreasing sanitary risks.

5.3.2.4 A Giant Urban Market Right at the Farm's Gate

Egyptian consumers appreciate milk with a high fat content. Yet, 37% of our sample, especially the families with the smallest assets, mentioned that they were not able to regularly sell milk, because of low production. If the family produced enough milk to create a surplus, direct sales to consumers at the farm's gate was the most common practice. In 2013, the average consumer price for buffalo milk was around €0.73 per litre with variations depending on the area investigated and the general income level of the population. It was not possible to properly identify a strong community of cow's milk producers within the city. It wasn't commercialized as an individual product, but was mixed with buffalo milk and sold for €0.63 per litre. Although some farmers admitted mixing buffalo and cow's milk, others used cow's milk surreptitiously to increase their volume without informing their customers. Only a small number of farmers sold their milk through intermediaries (17% of the sample). In these cases it was generally due to a social or family relationship with the dealer. Six families, with more significant assets, started to introduce processed milk products, but the practice remained marginal in our sample. They sold some yogurts products, white cheeses and butter in winter, and during feasts (Eïd, Ramadan, etc.) when dairy product prices tend to rise for a short time. To sell these products, some invested in small, minimally-equipped milk shops that are commonly found in all popular areas of Cairo. This marketing practice seems to be a way to increase the profits of their milk production, but according to many of those interviewed, a large number of customers, mainly women, still prefer to buy raw buffalo milk and make their own dairy products, each family having its own recipes.

The increasing presence of agro-business companies, promoting standardized products, is beginning to create new demands from consumers for products that have been tested for quality. Among the farms investigated, no conventional milk quality tests were applied. An organoleptic assessment, family reputation and customer satisfaction seemed to be the main guarantee of quality.

Artisanal laboratories are well established in Cairo, but presently, no relevant data are available to estimate their number and diversity. They are processing milk into cheeses (white cheese, some mozzarella) and using different milks (raw or skimmed buffalo or cow's milk) from all over Egypt. We have not established any link between UPA in our sample and those processing units, nor with agribusinesses enterprises.

5.4 Conclusion

A rich diversity of producers acting in Greater Cairo has been identified. These farms are remaining by taking advantage of their urban environment and constantly adjusting their systems to the cities changes. Cairo's UPA has been adapting for decades, developing interactions with the city that have lead to a symbiotic relationship. The farms of Cairo's UPA system provide several services to society that support their continued presence within the megalopolis. Their main asset is their ability to produce a highly valued product: fresh, raw buffalo milk. This production within the city combines three advantages. First, like in Iran, Azerbaijan and many other countries, buffalo milk is extremely sought after by Egyptian consumers

(Borghese and Mazzi 2005). This demand is seen at all levels of income, and stems from an appreciation for its high fat content and traditional use in many recipes. The second asset is the ability of those producers to sell truly fresh milk, *directly from the buffalo*. Consumers prefer a very short supply chain, to help guarantee the quality of a fresh product, and avoid fraud. As a third advantage, UPA provides a geographically, accessible point of sale. In a city like Cairo, with strong mobility limitations, close proximity represents a strong asset by limiting the distance consumers must travel to obtain the product.

Besides a milk supply, Cairo's UPA provides food security to a number of the city's inhabitants with limited purchasing power. Food produced is often shared between close or extended family members that are able to benefit from arrangements outside of monetary transactions. UPA also provides employment (Padgham et al. 2015). Producers, inputs providers, and intermediaries represent a segment of citizens who often lack the qualifications to find work in a saturated employment market. Finally, even if the phenomenon is hard to control and to estimate, the benefits of UPA ensure at least some agricultural land protection when dedicated institutions have limited control on encroachment (Allen et al. 2014).

The regulation of UPA activity in Cairo is very limited. Since neither the state, the police nor the military have much control on informal settlements (El-Batran et Arandel 1998), social capital appears to be the primary element organizing these areas (Belge and Blaydes 2014). Thus the challenge for the local producers is to be accepted and recognized by their neighbors, much as we see in many other cities in Africa (Robineau 2013; Padgham et al. 2015). Inside each zone, families have strong social links tying them together. A long common history between farmers and their neighboring consumers, often mixed with family bonds facilitates their tenacious presence within Cairo. With family reputation and trust playing a major role in the guarantee of product quality, the close proximity of producers becomes a benefit for the consumer.

Besides this core social capital, the adaptation capacity of these systems appears to be crucial. In the context of Greater Cairo, agricultural adaptation mechanisms generally appear to be initiated at the family level. If no "formal" channel promotes new practices within the community, farmers succeed by developing adaptive strategies and activating new resources within their environment (Pecqueur 2005). Indeed, the city acts as a compelling factor, pushing farming systems to evolve but also offering new resources in return. The dense social network of interactions between farmers and suppliers seems to be the major framework in which practices are designed, promoted, modified, and adjusted in a dynamic and constant process. The heterogeneity of the population in Egypt's capital city, with its large variety of profiles and increasing opportunities for young people to use new technologies, contributes to a fertile melting pot where the proximity of actors strengthens the adaptive process. In Cairo, the feeding systems appear to be one of the most dynamic areas where farmers may be able to increase their farming system's efficiency. As seen in studies of the capital city of Uganda, Kampala (Sabiiti et al. 2014), these feeding adaptations provide an innovative opportunity to recycle a part of the city's organic wastes. Besides milk, manure production in Egypt becomes a precious activity as the country tries to extend its agricultural land into the desert. This underlines the assessments that suggest UPA's role in ecosystems services needs to be better quantified (reuse of wastes, manure production) (Padgham et al. 2015). But a lack of structuration of the urban farmer's community, from the point of view of an adaptive process, highlights the difficulty of promoting new practices beyond the extended family or neighbourhood.

Although the capacity of Cairo's dairy farmers to create innovative adaptations has been established, their ultimate sustainability in an environment under constantly increasing urban pressure remains questionable. In terms of production, the lack of available space seems to be the most threatening factor. Even though input prices, especially feed, may increase in the future and challenge profitability, the demand for dairy products may equally continue to grow. For the moment, consumers seem to be strongly attached to traditional products made with buffalo milk and appear ready to pay higher prices for it, but it is not certain that this condition will remain the same in a few years. No connection between Cairo's UPA and the rapidly expanding agribusiness industry has been identified, but it could be considered an indirect and innovative driver of the dairy sector. Indeed there are factors which may influence future generations to avoid buffalo milk: An aggressive marketing campaign from agribusiness industry, portraying the artisanal dairy sector as a dangerous and unhealthy supply channel, the increasing presence of supermarkets, and the recognition of obesity as a national health issue. Under these influences, more economical standardized products made with fresh or powdered cow's milk with a lower fat content may be the emerging market. Although a few producers have been shifting from selling raw buffalo milk directly to selling processed products, their ability to prove the quality of their products with factual tests remains uncertain but may be an important demand from the coming generation of consumers. Finally from a governmental point of view, public health could be threatened by these systems if they cannot verify product safety information, and the current tolerance of the farm system's neighbours may decrease as the contemporary urban lifestyle becomes more prevalent, increasing the pressure for these farms to leave or abandon their production.

Besides contributing to the food security of Cairo's inhabitants, UPA provides many services, including employment, manure production, and wastes recycling. These are elements in the urban system that will have to be taken into account in the sorely needed efforts in urban planning. Even if these families appear extremely dynamic in their practices, successfully integrating resources that the city provides, it doesn't change the fact that their future appears uncertain in the increasingly competitive market of Greater Cairo.

Acknowledgments We would like to thank all the farmers, their families and intermediaries who have participated in the fieldwork and made this research possible at a very particular time in Egypt. Our thanks also go to the APRI (Animal Production Research Institute) in Dokki, for their expert advices, and certainly to the AIRD young team program for financing this project. Finally, thanks go to all the CIRAD-SELMET team for their precious support.

References

Adriansen HK (2009) Land reclamation in Egypt: a study of life in the new lands. Geoforum 40(4):664–674
Allen A, Frediani AA, Wood-Hill M (2014) 13. Land and planning for urban agriculture in Accra: sustained urban agriculture or sustainable urbanization? In: Irrigated urban vegetable production in Ghana, pp 161
Alary V, Corniaux C, Aboulnaga A, Galal S (eds) (2016) The traditional milk sector around Greater Cairo in Egypt. CIRAD, APRI, ICARDA, Cairo University, Ain Shams University, Montpellier. https://umr-selmet.cirad.fr/FichiersComplementaires/Atlas_finale_BR1_end.pdf
Anderberg MR (1973) Cluster analysis for applications. Academic, New York
Belge C, Blaydes L (2014) Social capital and dispute resolution in informal areas of Cairo and Istanbul. Stud Comp Int Dev 49(4):448–476
Borghese A, Mazzi M (2005) Buffalo population and strategies in the world. Buffalo Prod Res 67:1–39
Brossier J (1987) Système et système de production. Cah Sci Hum 23
Daburon A, Tourrand JF, Alary V, Ali A, Elsorougy M (2014) Typologie exploratoire des systèmes d'élevage laitier familiaux de la mégalopole du Grand Caire en Egypte, contraintes et opportunités. Rev Elev Med Vet Pays Trop 67(4):173–182
Dixon M (2014) Desert development and agribusiness in Egypt: making land (and money) a commodity, Watch letter No. 28, p 87. CIHEAM
Dray S, Dufour A-B (2007) The ade4 package: implementing the duality diagram for ecologists. J Stat Softw 22(4):1–20
El-Batran M, Arandel C (1998) A shelter of their own: informal settlement expansion in greater Cairo and government responses. Environ Urban 10(1):217–232
El-Hefnawi AIK (2005) Protecting'agricultural land from urbanization or 'managing'the conflict between informal urban growth while meeting the demands of the communities (lessons learnt from the Egyptian policy reforms). In: Third urban research symposium on land development. Urban Policy and Poverty Reduction, Brasilia, Brazil. World Bank Institute of Applied Economic Research (IPEA) Brasilia, DF, Brazil
El-Nahrawy MA (2011) Country pasture/forage resource profiles. Food and Agriculture Organization of the United Nations (FAO)
Goodman L (1961) Snowball sampling. Ann Math Stat 32(1):148–170
Gyasi EA, Fosu M, Kranjac-Berisavljevic G, Mensah AM, Obeng F, Yiran GAB, Fuseini I (2014) Building urban resilience: assessing urban and Peri-urban agriculture in tamale. Ghana
Hill MO, Smith JE (1976) Principal component analysis of taxonomic data with multi-state discrete characters. Taxon 25(2/3):249–255
Johannsen AT, Mahrous MN, Graversen M (2009) Landownership disputes in Egypt-a case study of the tensions around the Monastery of Abu Fana in May 2008. Cairo, p 63
Kipper R, Fischer M (2009) Cairo's informal areas between urban challenges and hidden potentials. GTZ Egypt, Cairo
Landais E (1996) Typologies d'exploitations agricoles. Nouvelles questions, nouvelles méthodes. Economie Rural 236:3–15
Padgham J, Jabbour J, Dietrich K (2015) Managing change and building resilience: a multi-stressor analysis of urban and peri-urban agriculture in Africa and Asia. Urban Clim 12:183–204
Pecqueur B (2005) Le développement territorial: une nouvelle approche des processus de développement pour les économies du Sud. In: Antheaume B, Giraud F (eds) Le territoire est mort. Vive les territoires! IRD, Paris
Popkin B, Wen Ng S (2007) The nutrition transition in high-and low-income countries: what are the policy lessons? Agric Econ 37(s1):199–211
Robineau O (2013) Vivre de l'agriculture dans la ville africaine: une géographie des arrangements entre acteurs à Bobo-Dioulasso. Université Paul Valéry-Montpellier III, Burkina Faso

Sabiiti EN, Katongole CB, Katuromunda S, Sengendo H, Basalirwa CPK, Atukunda G, Nambuubi SK (2014) Building urban resilience: assessing urban and Peri-urban agriculture in Kampala, Uganda

Séjourné M (2006) Les politiques récentes de « traitement » des quartiers illégaux au Caire: nouveaux enjeux et configuration du système d'acteurs ? Université François Rabelais de Tours, Tours

Warren E, Hawkesworth S, Knai C (2015) Investigating the association between urban agriculture and food security, dietary diversity, and nutritional status: a systematic literature review. Food Policy 53:54–66

Part II
Methods and Tools to Design Sustainable Urban Food Systems

Chapter 6
An Ecological Footprint-Based Spatial Zoning Approach for Sustainable Metropolitan Agro-Food Systems

Dirk Wascher, Leonne Jeurissen, Jan-Eelco Jansma, and Michiel van Eupen

Abstract Based on research undertaken in Berlin, Milan, London, Ljubljana, Rotterdam and Nairobi, the EU FP7 Project 'Food Planning and Innovation for Sustainable Metropolitan Regions' (www.foodmetres.eu) has developed a string of assessment tools and food planning concepts that are especially geared towards the role of metropolitan regions as spatial-functional entities of sustainable food supply. Focusing on the example of the Rotterdam Metropolitan Region, this paper puts forward a spatial-analytical approach that combines an ecological footprint analysis with land use assessment and knowledge on the impacts of food chains on the social, environmental and economic dimensions of the agro-food sector. Drawing upon existing work, we have developed an approach to delineate Metropolitan Agro-Food Systems (MAS) as spatially explicit foodsheds. Taking the food demand of a city into consideration, the required amount and location of 'local hectares' of agricultural areas meeting these demands are identified as the starting point illustrating the challenge of feeding an urban population. Using the example of the Rotterdam region, we show how the von Thünen model can be adapted to a dynamic interpretation of the regional food supply potentials on the basis of local hectare calculations deriving from food consumption census data.

Keywords Spatial analysis · Urban food consumption · Food planning

---

D. Wascher (✉) · L. Jeurissen · M. van Eupen
Wageningen Environmental Research, Wageningen University and Research, Wageningen, The Netherlands
e-mail: dirk.wascher@wur.nl; leonne.jeurissen@wur.nl; michiel.vaneupen@wur.nl

J.-E. Jansma
Applied Plant Research, Wageningen University and Research, Wageningen, The Netherlands
e-mail: janeelco.jansma@wur.nl

C.-T. Soulard et al. (eds.), *Toward Sustainable Relations Between Agriculture and the City*, Urban Agriculture, https://doi.org/10.1007/978-3-319-71037-2_6

6.1 Introduction

The European Sustainable Development Strategy (SDS) addresses a broad range of 'unsustainable trends' ranging from public health, poverty and social exclusion to climate change, energy use and management of natural resources. A key objective of the SDS is to promote development that does not exceed the ecosystem's carrying capacity and to decouple economic growth from negative environmental impacts. A report commissioned by the European Commission (2008) came to the conclusion that the Ecological Footprint should be used by EU institutions within the Sustainable Development Indicators (SDI) framework.

The Ecological Footprint measures how much biologically productive land and water area is required to provide the resources consumed and absorb the waste generated by a human population, taking into account the prevailing technology. The annual production of biologically-provided resources is called bio-capacity. The Ecological Footprint and bio-capacity are each measured in global hectares, a standardized unit of measurement equal to 1 hectare with global average productivity (European Commission 2008).

However, due to a fragmented research history with simultaneous and largely uncoordinated efforts across sectors, research institutes and regions, ecological footprint calculations are manifold and differ substantially in terms of underlying data and methodologies (Verzandvoort et al. 2013). While the ecological footprint is still considered as a key reference and communication tool when comparing environmental impacts at highly aggregated levels, the above-mentioned inconsistencies have been a matter of concern for both research and policy.

Another challenge of the ecological footprint approach is the abstract dimension of global hectares which represent the total impact of certain economic sectors and activities as the sum of all processes along the production chain – in this case the food chain from farm to fork. This includes all energy, water, land and material input resources such as fertilizers, machinery and packing material that occur along the full food chain. Global hectares can be considered a successful communication tool that addresses the complexity of the life-cycle analysis by means of spatial impacts on land demand.

At the same time, the virtual nature of the resulting impact figures and graphs (land demand as square kilometres depicted as generic circles around impact sources, e.g. city centres) provide images without any geographic, socio-cultural context: global hectares treat the world as a uniform landmass that is being consumed by the impacting human activities, i.e. food consumption. Because global hectares include the full range of impacts along the whole food chain, resulting figures – certainly for large metropolis – tend to be of substantial size (see Fig. 6.1). However, coinciding with or in response to the use of 'global hectares' for communicating impact dimensions, there appears to be an increasing demand for developing sustainable food security strategies in which the regional supply coming from agricultural areas in and around cities is playing a central role. The question is not any longer 'how big is the impact?', but 'how can we lower this impact?' and 'how

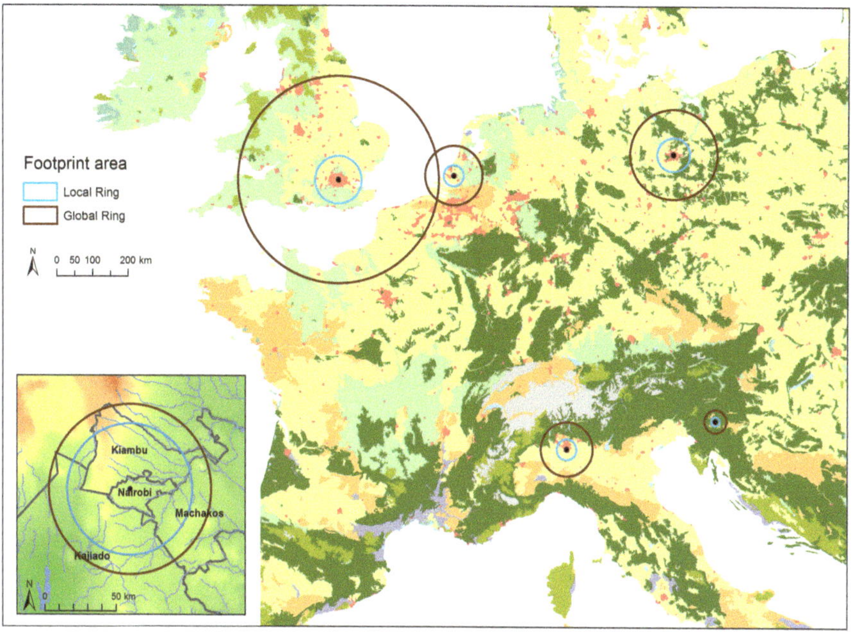

Fig. 6.1 Ecological footprint (EF) shown as global and local hectares for London, Rotterdam City Region, Berlin, Milan, Ljubljana and Nairobi. Large dark circles refer to global hectares and small blue circles to local hectares showing the land requirements in terms of food production areas based on national accounts (Sources: global hectares according to Briggs 2011; local hectares various sources, see below)

can we feed our cities in more sustainable ways?'. Given that a large proportion of food consumption impacts are related to transport and logistics associated with global food networks, increasing the regional food supply has been heralded as both a social as well as an environmental asset (EEA 2014; Steen-Olsen et al. 2012). So-called 'Short Food Supply Chains' are considered to substantially decrease CO_2 emissions deriving from food transport and increasing the diversity and quantity of regional food supply will most likely stimulate regional economics allowing for more social cohesion, work opportunities, fair competition, innovation and specialisations along the food chains – in short a re-vitalisation of agro-food systems in the metropolitan regions of cities (Kneafsey et al. 2013).

Here is where the Metropolitan Foodscape Planning (MFP) tool comes in. Rather than relying on global hectares as the basis for communicating impacts, this tool focuses on identifying 'local hectares' – namely concrete agricultural land use areas needed for producing those regionally growing food crops that are required to feed urban populations. The development of this tool has been guided by the following research questions:

- What can be considered as sustainable boundaries of a metropolitan food shed?
- What is the agricultural resource potential of such metropolitan food sheds?

- To which extent does it cover the metropolitan food demand?
- Which type of policy and planning instruments can help to reduce the impact of food consumption on the ecological footprint?

This paper first addresses the need for widening the concept of the metropolitan region towards an agricultural resource base that can be considered as the regional food shed and driver for agricultural innovation. We then explain why the concept of a metropolitan food shed requires a spatial planning dimension to ensure that the regional food production is integrated with other spatial demands such as nature conservation, recreation and health aspects. Using the example of the Rotterdam region, we present the technical application of the MFP-tool in terms of map products and food surplus/deficit graphs for a total of nine food groups. Following a discussion of these results, we draw a series of conclusions addressing food planning issues in metropolitan regions.

6.2 Widening the Concept of Metropolitan

The attempt of defining – let alone systematically framing – a metropolitan region meets a series of methodological and conceptual challenges. The concept of metropolitan regions has its origins in the fast growing urban development processes of North America at the end of the nineteenth century and only recently of Asia. Given the speed and dynamics of these urban developments, terms such as '(mega-)city', 'suburban' or 'urban agglomeration' do not appear as appropriate since they think urban structures as being isolated from their surroundings. These areas around dense and expanding urban settlements are part of urban processes by developing symbiotic relationships and characteristics that are worth recognizing (Smeets et al. 2004; Smeets 2009; van Steekelenburg 2009; van Eupen et al. 2012). Within only a few years – in China, nowadays, even within months – the shape and extension of cities can dramatically change, transforming former rural areas into quasi-urban structures, linking formerly fragmented areas by expanding transport networks while simultaneously facilitating daily job commuting and recreational-cultural activities during weekends (Taylor 2012). Because of these socio-economic origins, the notion of metropolitan regions is frequently associated with urban-regional planning concepts for effectively managing transport, housing, recreation as well as industrial and commercial development based on administrative regions (see Fig. 6.2). Reflecting such a view, the goal of the Metropool Regio DenHaag Rotterdam (MRDR) is to strengthen its economic competitiveness and its attractiveness as a business location. At the same time the region is meant to satisfy recreational needs and offer experiences in nature (MRDH 2014). The introduction of parkways and large urban green zones in American cities such as Baltimore, Chicago and New York by landscape architect Frederic Law Olmsted (MacDonald 2013) integrated natural elements at a rather early stage of the metropolitan development (Birnbaum and Hughes 2005). Today, the concept of 'green lungs' and

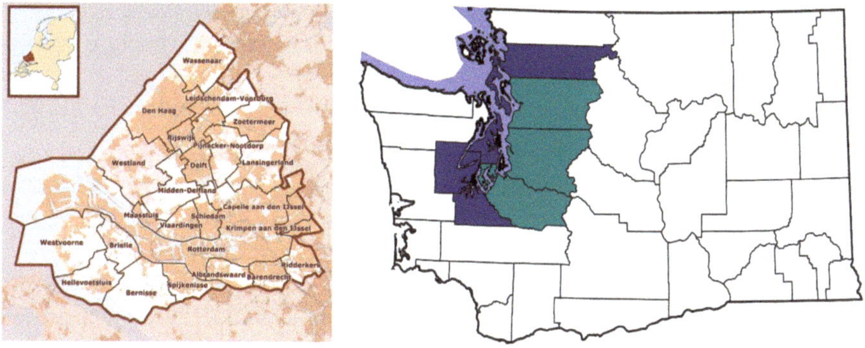

Fig. 6.2 Bangkok Metropolitan Area (Thailand), Metropool Regio DenHaag-Rotterdam (MRDR) in The Netherlands and Seattle Metropolitan Areas (US)

'green fingers' (corridors) have become assets of urban development plans all over the world (Antrop 2004; Caspersen et al. 2006).

Following a series of scientific seminars among geographers and landscape researchers, Tress et al. (2004) differentiated between four categories of metropolitan landscape definitions, namely as (1) a synonym for urban areas, (2) an agglomeration or administrative city region, (3) a large supra-regional area, and as (4) a space of continuous urban influence. It is interesting to note that none of the accompanying text references that elaborate these definitions further mentions 'agricultural land use'.

And indeed, for a long time, agricultural land use did not seem to be part of most contemporary metropolitan concepts – even among European geographers and landscape architects. It is only recently – and strongly driven by Dutch planners, researchers and agronomists – that urban demand, lifestyle and business are considered 'game-changers' with regard to the notion of rurality, agricultural supply and landscape character near cities (Van Latesteijn 2008; Tress et al. 2004; Smeets 2009; Wascher et al. 2010).

6.3 A Spatial Zoning Concept for Linking Food Demand and Supply

Expanding on the concept of agricultural supply, the Dutch think-tank Transforum specifies metropolitan agriculture as "a deliberately designed system of intelligently connected [agricultural] production sites that uses the available resources, conditions and infrastructure in metropolitan areas to produce material and immaterial demands for the same metropolitan area" (van Latesteijn 2008). This description suggests:

1. Spatial-functional entities with boundaries which are determined by system integration at the production level, thereby defining what constitutes to a metropolitan area;

2. Sustainable principles, among them the limitation of agriculture's ecological footprint by the improved use of resources, conditions and infrastructure that are available in the area of demand;
3. A multifunctional approach by covering society's material as well as immaterial demands (commodity and non-commodity goods and services).

The above interpretation links well with the goal set for the Metropool Regio Den Haag Rotterdam (MRDH 2014). However, it also becomes obvious that administrative regions are of limited value if it comes to the question of food security and the supply functions of landscapes with regard to urban food demand: land use, soil quality and agricultural supply capacities are hardly guided by administrative boundaries, but follow bio-physical and entrepreneurial criteria.

Translating the vision of modern metropolitan agriculture into a spatially explicit planning concept for food security requires a more dynamic approach that is based on multi-functionality, evidence-based planning principles and multi-actor governance.

One early example for a food security planning concept dates back to the early nineteenth century: Johann Heinrich von Thünen (1783–1850) developed a theoretical model describing how market processes determine local land-use patterns. Addressing the historical economy of his time (1826), von Thünen considered dairying and horticultural farming – given their value and perishability – to be located closest to the city. Timber and firewood – because of its weight – should be produced in the second zone. The third zone consists of extensive field crops such as grain for bread. Livestock rearing was reserved for the final ring because they are self-transporting and thus have low transport costs (see Fig. 6.3).

It is not surprising that the von Thünen approach – despite its continuing popularity – has been criticized for being too rigorous, too static and in terms of its geopolitical setting simply out of time. However, the model's focus on agricultural supply in the light of urban food demand as well at its two key indicators, namely land prices and transportation costs, are still valid today.

Inspired by the von Thünen principles, we established the rationale that urban core areas should be surrounded by (1) a green belt reserved for experiencing nature and recreational activities, followed by zones offering a metropolitan food supply such as (2) an intensive farming zone for products finalized for plant-based diets, followed by (3) a zone entirely dedicated to urban needs for animal-protein-based food, and (4) a transition zone providing food products for both plant-based diets as well as animal proteins for urban populations within the wider metropolitan territory. Zones 2 and 3 are considered to fully satisfy the yearly food demand of the urban core area – in the case of Rotterdam, the 1.2 million citizens of the Rotterdam City Region. The transition zone 4 is meant to provide sufficient food for the wider metropolitan area making use of the OECD approach which includes other cities such as Den Haag, Delft or Dordrecht with a total of 6.6 million citizens.

With regard to the goal to spatially and functionally define Metropolitan Agro-Food Systems (MAS) as highly multi-functional food sheds of urban core zones, we combine two conceptual approaches:

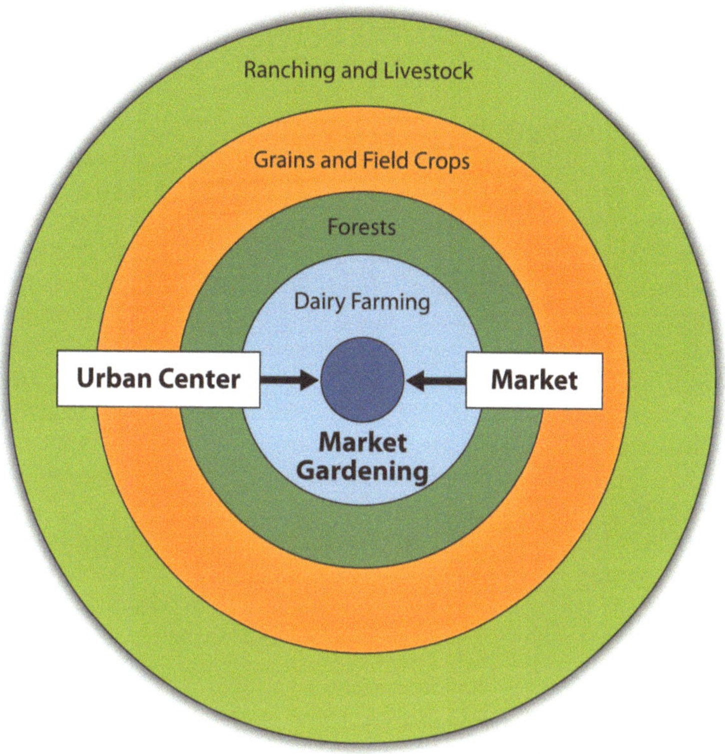

Fig. 6.3 von Thünen's model: the central dot represents a city; dairy and market gardening; forests for fuel; grains and field crops; ranching; the area outside these rings represents wilderness where agriculture is not profitable (von Thünen 1826)

- References developed by geographical and planning analyses such the characterisation of urban sprawl (Glaeser and Kahn 2003; Deng et al. 2010), central place theory (Christaller 1933), accessibility (Alonso 1964; Litman 2003; Halden et al. 2005), mobility and transports (Wascher 2012); and
- The production capacity of agricultural land to satisfy all, or the majority of the urban food demand (Chen et al. 2010; Wackernagel and Ress 1996; Gerbens-Leens et al. 2002). This capacity varies according to several factors, such as food products, seasonality, soil quality and yield, the latter depending also on productive inputs and specific agro-climatic variables.

In the following we will introduce the Metropolitan Foodscape Planning tool that allows to (1) spatially define MAS as a territorial approach delineating the actual food supply on the basis of urban food consumption, and to (2) provide a spatial zonation for both communicating the impact (= footprint) as well as for guiding future land use decisions taking into account aspects such as recreation, nature conservation and food safety.

6.4 Metropolitan Foodscape Planning Tool: Methods and Data

The Metropolitan Foodscape Planning (MFP) tool has largely been inspired by the work undertaken by Jansma et al. (2012; Jansma 2013) when developing the Urban Foodprint Tool, an internet-based assessment tool that offers local footprint calculations (in terms of ha or acres to feed an certain amount of people) on the basis of food productions and diet in the Netherlands (www.stedelijkefoodprint.nl/51-urban-foodprint.html). However, rather than producing quantitative outputs (e.g. hectare requirements for certain food groups based on food consumption figures for The Netherlands), the MFP tool was meant to also show the extend and location of the actual supply with these required food groups in a geographical context. Designed as a European assessment tool, the MFP also needed to take into account the different national diets of the other participating countries (United Kingdom, Slovenia, Germany and Italy). For this purpose we made use of European sources (Best et al. 2008; EFSA 2011) in combination with national assessments.

Building the Metropolitan Foodscape Planner tool (MFP) requires hence a series of data management and GIS operations to be performed in Excel and Arc-Info (Table 6.1). Using the example of the Rotterdam Metropolitan Region procedure, we will illustrate the following sequence of steps that are required:

- Assessing the ecological footprint for food consumption
- Creating the dynamic footprint-driven spatial zoning framework (von Thünen);

Table 6.1 Data layers applied in the MFP model

Data layer	Source
Corine Land Cover 2006	http://www.eea.europa.eu/data-and-maps/data/corine-land-cover-2006-raster-3
	Version 8 April 2014, download 13 Jan 2015
	In arccat export .tiff als esrigrid in MFT.gdb
Protected areas (Natura2000)	http://www.eea.europa.eu/data-and-maps/data/natura-5#tab-gis-datashapefileNatura2000_end2013_rev1.shp
Landscape units (Lanmap2v1)	European Landscape Typology (LANMAP) (Mücher et al. 2010) lanmap2_v1_level_4_ls-cod
Food-rings based on food demand in ha multiplied by population numbers	Food demand: EFSA (2011), Brabantse Milieufederatie (2009), Jansma et al. (2012) and Jansma 2013
	Population numbers: City of Rotterdam
	OECD metropolitan region: Brezzi et al. 2012
HSMU	Homogenous Soil Mapping Units (HSMU) as modelled by CAPRI (Kempen et al. 2005) and Eurostat crop area data;
	Year per country: NL 2008, BL 2008, DE 2008, PL 2004.

- Disaggregating the CORINE land cover units to arrive at distinctive land use types in form of commodity groups (HSMU);
- Establishing commodity group allocation rules on the basis of landscape units based on LANMAP (Mücher et al. 2010).

The actual identification of the food-supply zones followed the step-wise approach described below.

6.4.1 Urban Core and Green Buffer

While the size of all food-production zones corresponds to the land demand deriving from the urban food demand (total population figure × local hectare land demand per capita), the designation of the Green Buffer takes on an expert-based approach assuming a spatial relationship between the urban settlement area and the projected need for sufficient green space in close distance. In the case of Rotterdam, the radius of the Urban Core is 10 km (Table 6.2). For the Green Buffer half that distance – thus 5 km – has been taken. Within this Green Buffer we did not consider existing land use areas to be eligible for land use change/food group allocation plans. We did though consider to maintain existing grasslands to contribute to extensive livestock farming as in the past. Since extensive livestock farming appeared to be acceptable or even desirable in the traditional landscapes in close proximity to cities, we took the available grassland (14,463 ha) within the urban core and the green buffer to calculate how many people can be fed on animal protein-based diets from Zone 1 (288,720 persons).

6.4.2 Two Metropolitan Food Rings

The radii of Zone 2 (crops for plant-based food) and Zone 3 (crops for livestock-farming) are calculated based on the total demand in hectares for the population and the total area available for agriculture per ring. For the Rotterdam city region the population for the Metro-Food-Ring is 1.2 million. Table 6.3 shows the demands we used for calculating the food supply zones. Preparatory work for calculating the food supply zones include the conversion of vector data (Corine LC, Lanmap & Natura 2000) to 1 ha raster data, and the adaptation of the projections and several re-classifications to separate agricultural from non-agricultural land use types (grey colour). Zone 2 (crops for plant-based food) is based on the city's population multiplied by the demand factor for conventional plant-based diets (0.0341 ha/person), using total available land that is "arable, but not grass" to calculate the required size of this zone (41,129 ha). Zone 3 (crops for livestock farming) is based on the Rotterdam Cityregion population minus 'Zone 1 (animal protein consumers) multiplied by the demand factor for conventional animal products (0.178 ha/person) to calculate the required size (163,445 ha).

Table 6.2 Calculations for the MFP zoning distances for the City Region and OECD Region of Rotterdam

Ring types (zones)	Distance (km)	Surface area of rings (ha)			Demand factor (ha/p)			Population	Required surface area (population × demand factor)	Land use type
		Arable or grass	Arable	Grass (includ. protected)	Arable or grass	Arable	Grassland			
Rotterdam city region								1,200,000		
Rotterdam OECD region								7,800,000		
Urban Core	0–10	12,642								
Green Buffer	10–15	25,608								
Organic dairy in Urban Core and Green Buffer	0–15	38,250		14,436			0.05	288,720		Grass, irrespective of protection
Metro-Food-Ring (plant-based)	15–24	68,930	41,129			0.0341		1,200,000	40,920	Arable, not protected
Metro-Food-Ring (protein-based)	24–40	163,445					0.178	911,280	162,208	Arable and grass, not protected
Transition-Zone	40–150	1,402,085			0.2121			6,600,000	1,399,860	Arable and grass, not protected

See data sources in Table 6.1 and data breakdown for demand figures in Table 6.3

Table 6.3 Breakdown of supply and demand based on CORINE LC – HSMU disaggregation for all three MFP-zones of Rotterdam City Region & OECD region

Population	1,200,000				911,280				6,600,000			
Zone	Zone 2 (crops for plant-based food)				Zone 3 (crops for livestock farming)				Zone 4 (transition)			
Food groups	Ha/ capita	Demand (ha)	Supply (ha)	Surplus/ deficit (ha)	Ha/ capita	Demand (ha)	Supply (ha)	Surplus/ deficit (ha)	Ha/ capita	Demand (ha)	Supply (ha)	Surplus/ deficit (ha)
Crop rotation	0.0163	19,560	18,794	−766		0	35,677	35,677	0.0163	107,580	309,528	201,948
Other cereals	0.0030	3600	1395	−2204		0	5844	5844	0.0030	19,800	113,597	93,797
Oilseed plants	0.0003	360	105	−255		0	600	600	0.0003	1980	4678	2698
Fodder		0	5920	5920	0.1280	116,644	18,664	−97,980	0.1280	844,800	372,776	−472,024
Vegetables	0.0015	1800	4984	3184		0	20,207	20,207	0.0015	9900	59,349	49,449
Fruit	0.0130	15,600	526	−15,074		0	1567	1567	0.0130	85,800	26,852	−58,948
Wine		0	9	9		0	13	13		0	53	53
Grassland		0	37,195	37,195	0.0500	45,564	80,873	35,309	0.0500	330,000	515,252	185,252
Total	0.0341	40,920	68,930	28,010	0.1780	162,208	163,445	1237	0.2121	1,399,860	1,402,085	2225

Based on Jansma (2013) and Brabantse Milieufederatie (2009)

6.4.3 Transition Zone

Zone 4 is meant to feed the wider metropolitan region according to the OECD (Brezzi et al. 2012), which included an additional 6.6 million people living in cities such as Den Haag, Delft and Doordrecht. As its title 'transition zone' indicates, these areas of land are to be shared with adjacent metropoles overlapping with the transition zones of other nearby cities such as Antwerp or Amsterdam. The transition zone integrates crops for both plant-based food and livestock farming, accounting for 0.21 ha/person to calculate the required size of this zone (1.4 million ha).

Identifying crop types in terms of their geo-referenced locations has been the most challenging step when constructing the MFP tool. Corine Land Cover data only offers highly aggregated crop-unspecific land cover types. National land use data on the contrary varies strongly between countries and does not allow for a European-wide approach. First efforts to make use of FAO GEAZ yield data (2010 at a 10 km-grid resolution did not lead to the type of crop representations that we deemed necessary to spatially allocate some of the key crop types. We hence settled for the Homogenous Soil Mapping Units as modelled by CAPRI (Kempen 2012) at the resolution of 1 ha raster size. Though less accurate than national land use survey data, HSMU is available for the whole of Europe, allowing direct top-down assessments without resource-consuming data gathering procedures.

HSMU encompasses the major crop types such as wheat, potatoes, sugar beet (which we grouped into a crop rotation unit), other cereals, vegetables, fruit, oilseeds, fodder crops, grasslands, rice, wine and others. Comparing the reference data (present situation as of 2009) with the crop-specific demand figures calculated on the basis of the regional footprint allows to compare demand with actual supply, resulting in surplus/deficit figures (see Table 6.3).

Figure 6.4 shows the results of applying the MFP tool to the metropolitan region of Rotterdam. The Green Buffer (Zone 1 at 10 to 15 km measured from the city centre) is thus positioned directly at the periphery of the urban core area of the Rotterdam municipal area. Table 6.3 informs that this zone includes about 25,600 ha of arable land and grassland. Being part of the Green Buffer, though, we propose that this land is reserved for extensive forms of urban agriculture and grassland farming, offering both food and recreational opportunities. Nature conservation areas and important ecological functions need to be preserved and managed accordingly.

Zone 2 (Fig. 6.5) is between 15 and 24 km distance from the centre and can be entirely dedicated to producing crops for plant-based food: all the consumption needs arising from the 1.2 million Rotterdam people can theoretically be satisfied within this zone. However, Fig. 6.5 shows that the current land use is still focusing strongly on livestock farming and that there are clear deficits for fruit (15,000 ha missing) and slight deficits for rotation crops, other cereals and oilseed plants. Exceptional, certainly when comparing to other European metropolitan areas, is the major surplus for vegetables (more than 3000 ha). This can be explained by the presence of the extensive areas of Dutch glasshouse production in Westland and

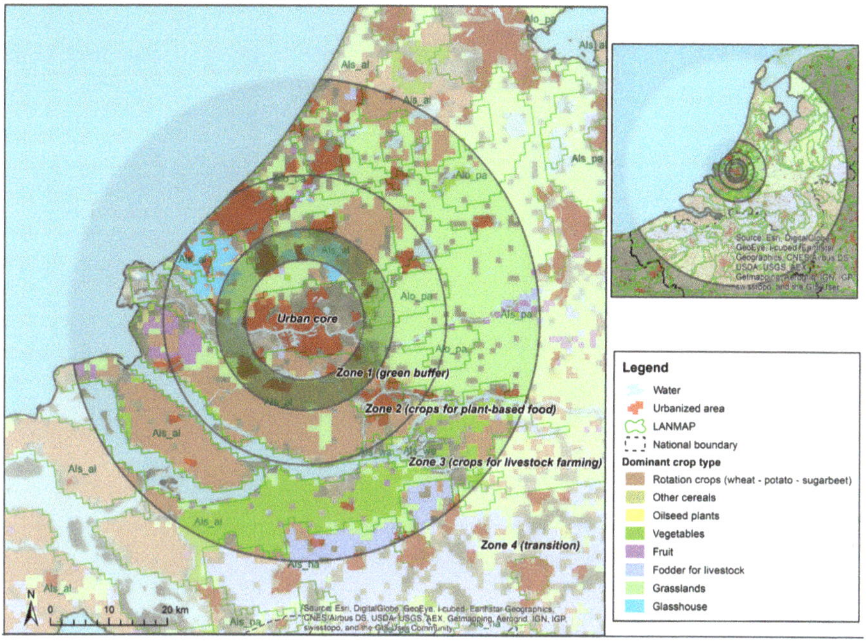

Fig. 6.4 Application of the Metropolitan Foodscape Planner to the Rotterdam City Region (1.2 million inhabitants) and the OECD metropolitan area (+6.6 million people) showing the current dominant crop type

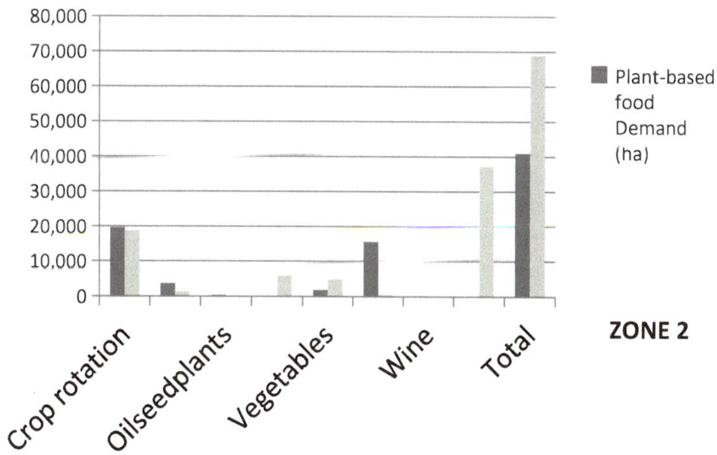

Fig. 6.5 Demand-Supply analysis for eight food groups of the Metropolitan Food Zone 2 (crops for plant-based food) for 1.2 million people (in hectares)

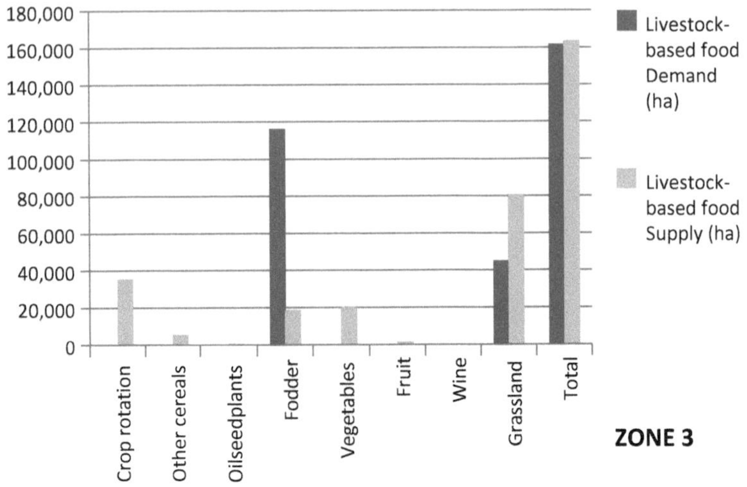

Fig. 6.6 Demand-Supply analysis for eight food groups of the Metropolitan Food Zone 3 (crops for livestock farming) for 1.2 million people (in hectares)

Oostland (see Fig. 6.5). Today most of this production (about 90%) is being exported and is strongly dominated by a few lead crops such as tomatoes, zucchini and bell pepper.

Zone 3 (Fig. 6.6) is between 24 and 40 km distance from the city centre. According to our scheme, this zone is entirely dedicated to crops supporting the city's demand for livestock such as dairy and meat products. Given the resource intensity of animal-based food products, it is not surprising that this zone requires a surface area four times as large than the one for plant-based food products in Zone 2: more than 160,000 ha. In this zone, the largest deficit is for fodder crops (almost 100,000 ha). Today these fodder crops are being imported from more remote Dutch locations and of course in the form of soya feedstuff from overseas amounting to about 20% of the total (van Gelder and Herder 2012). Morevover, we see a clear surplus of grassland production for dairy farming. In terms of the zone's diameter (16 km), it should be kept in mind that this is also a consequence of the city's location close to the North Sea where no land-based food production is possible.

Zone 4 (Fig. 6.7) spans over a distance from 40 km to 150 km from the city centre. This means that the Transition zone spans well into Belgium and Germany. Applying the OECD scheme as a reference (7.8 million people), this means that such a region covers almost half of the total Dutch population (16 million). Also here it is important to acknowledge the fact that Rotterdam's location at the sea means that agricultural land use is not available around the city and that supply only comes from the inland. As a consequence, the land requirement widen these inland circles considerably. Even so, the large area demands in terms of local hectares (almost 1.4 million) demonstrate the realities of densely populated regions here and elsewhere in the world. In terms of the demand-supply relationship, the transition

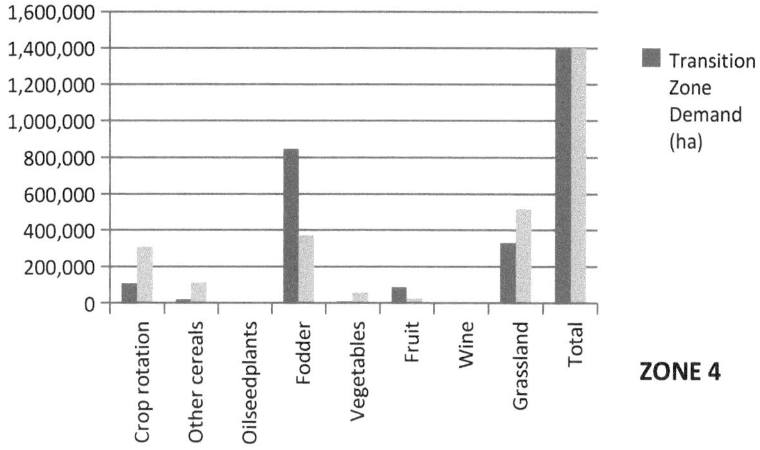

Fig. 6.7 Demand-Supply analysis for eight food groups of crops in Zone 4 (Transition) providing crops for plant-based food and livestock farming for 6.6 million people of the OECD region (in hectares)

zone mirrors the situation of Zone 3: the biggest deficit is for fodder crops required for livestock farming.

The MFP outputs presented in Figs. 6.5, 6.6 and 6.7 are not only meant to be assessment results for framing the impact of urban food production on the different metropolitan zones, but are also providing operational input to a stakeholder-oriented foodscape-planning device. For this purpose we introduce the data into the so-called 'digital maptable' which allows users to perform land use allocations by means of a digital pen. Addressing the surplus/demand figures resulting from the assessment, users can then make proposals for where and how to change the existing land use (food crops) in order to more properly meet the demands identified by the tool. Please see for further illustrations of the maptable approach Wascher et al. 2015.

6.5 Discussion and Conclusions

This project fills gaps in linking regional impact assessments to ecological footprint calculations, developing procedural spatial planning methodologies and offering process-oriented policy-science interfaces. The substantial part delivered building blocks for the role of sustainable food production and landscape planning and design.

Rather than serving as a normative model, the societal challenge of the MFP is to elaborate the spatial-functional link between land and food at both the community and the landscape level. The primary aim was to provide national, regional authorities as well as stakeholders with participatory tools to cooperate and contribute to the assessment and planning at the regional level; the secondary aim was to provide methodological feedback to European institutions' strategic policy goals on sustainable development.

Though the approach presented in this paper puts forward a strong focus on food production within the boundaries of a regional zoning system around cities, we would like to stress that the size of an ecological footprint depends on more factors than just travel distance to the city centre, namely yield factors depending on soils, climate and agricultural production techniques, e.g. tomatoes grown in Dutch glasshouses are of clearly less impact on water use than open field productions in many South-European locations. Ultimately, ecological footprint calculations need to take into account a far more complex system of environment-human interactions.

This also accounts for the way food consumption is being taken up by a footprint model. In order to keep the assessment manageable and to maintain a regional perspective, we have focussed only on those food groups/land use types which are currently present in the metropolitan regions of Rotterdam – or any other city which we researched within the FOODMETRES project. We hence did not address any exotic fruit or vegetables, no beverages like beer, wine or coffee and neither rice nor soya. Looking at some of the footprint literature (Chen et al. 2010; Wackernagel and Ress 1996; Gerbens-Leens et al. 2002) it is obvious that a regional supply without any major technological efforts (e.g. greenhouse productions) will not only result in a clearly less diverse daily menu, but will also be very dependent from the seasonality of these food groups. On the other hand, a recent report by the Dutch Economic Institute for Agriculture LEI (Terluin et al. 2013), came to the conclusion that the Dutch agricultural sector was found capable of supplying the country's population of 17 million inhabitants with a varied diet of both animal and crop products, providing each person with over 2000 calories a day. To achieve this, however, patterns of both agricultural production and food consumption would have to change substantially. According to the report, the most striking differences compared with a global market supply situation are the limited consumption of grain products (including bread), the complete absence of pork and the relative prominence of potatoes, chicken, and eggs in the diet.

However, we feel that the existing results demonstrate that quantitative assessments can provide important contributions to a wider societal debate on the characteristics and effectiveness of Metropolitan Agro-Food Systems (MAS). Using the example of the Rotterdam region, we show how the von Thünen model can be adapted to a dynamic interpretation of the regional food supply potentials on the basis of local hectare calculations deriving from food consumption census data. The results are considered as valuable references for the emerging discipline of sustainable food planning as a young branch at the very interface between social and environmental sciences. The project provides building blocks for interdisciplinary research crossing boundaries with the humanities (local identity, cultural landscape) and natural science (environmental impact, food technology). The initial applications of the 'maptable' device, when involving stakeholders during several workshops, have made the impact assessment capacities of the MFP-tool even more dynamic. Participants engaged commonly in fact-finding exercises when searching for new land use options when aiming for a reduced ecological footprint. In terms of further improving the tool in the future, adding the aspect of land prices, commodity market prices of food groups and transportation costs will certainly strengthen the tool's economic capacities.

References

Alonso W (1964) Location and land use. Harvard University Press, Cambridge

Antrop M (2004) Landscape change and urbanisation process in Europe. Landsc Urban Plan 67:9–26

Best A, Giljum S, Simmons C, Blobel D, Lewis K, Hammer M, Cavalieri S, Lutter S, Maguire C (2008) Potential of the ecological footprint for monitoring environmental impacts from natural resource use: analysis of the potential of the Ecological Footprint and related assessment tools for use in the EU's. Thematic strategy on the sustainable use of natural resources. Report to the European Commission, DG Environment

Birnbaum CA, Hughes MV (eds) (2005) Design with culture. Claiming American landscape heritage. University of Virginia Press, Charlottesville. 224 pages

Brabantse Milieufederatie (2009) How to feed Tilburg. Research implemented by Wageningen University with advice by Prof. Dr. Han Wiskerke, Wageningen. 33 pages

Brezzi M, Piacentini M, Rosina K, Sanchez-Serra D (2012) Redefining urban areas in OECD countries, in Redefining "urban": a new way to measure metropolitan areas. OECD

Briggs J (2011) The EUREAPA technical report. Stockholm Environment Institute, University of York. UK. One Planet Economy Network. 38 pages

Caspersen OH, Konijnendijk CC, Olafsson AA (2006) Green space planning and land use: an assessment of urban regional and green structure planning in Greater Copenhagen. Dan J Geogr 106(2):7–20

Chen D, Gao W, Chen Y, Qiao Z (2010) Ecological footprint analysis of food consumption of rural residents in China in the latest 30 years. Agric Agric Sci Procedia 1(2010):106–115

Christaller W (1933) Die zentralen Orte in Süddeutschland. Jena: Gustav Fischer (Translated (in part), by Charlisle W. Baskin, as Central Places in Southern Germany. Prentice Hall 1966)

Deng X, Huang J, Rozelle S, Uchida E (2010) Economic growth and the expansion of urban land in China. Urban Stud 47(4):813–843

EEA (2014) Environmental indicator report 2014. Environmental impacts of production-consumption systems in europe. The European Environment Agency, Copenhagen. 125 pages

EFSA (2011) Chronic food consumption statistics. Available at: http://www.efsa.europa.eu/en/datexfoodcdb/datexfooddb.htm

European Commission (2008) Potential of the ecological footprint for monitoring environmental impacts from natural resource use analysis of the potential of the Ecological Footprint and related assessment tools for use in the EU's Thematic Strategy on the Sustainable. Use of Natural Resources. DG Environment – Final Report. 312 pages

FAO (2010) FAO statistical yearbook. FAO, Rome, Italy, 2010. Available at: http://www.fao.org/docrep/015/am081m/am081m00.htm

Gerben-Leens PW, Nonhebel S, Ivens WPMF (2002) A method to determine land requirement relating to food consumption patterns. Agric Ecosyst Environ 90:47–58

Glaeser EL, Kahn ME (2003) Sprawl and urban growth. In: Henderson JV, Thisse JF (eds) Handbook of regional and urban economics. Elsevier, Amsterdam, pp 2481–2527

Halden D, Jones P, Wixey S (2005) Measuring accessibility as experienced by different socially disadvantaged groups. Accessibility analysis literature review working paper 3. Accessed from www.tecmagazine.com/universities

Jansma J-E (2013) Urban footprint manual. Assumptions, background and methods. http://www.stedelijkefoodprint.nl

Jansma J-E, Sukkel W, Stilma ESC, van Oost AC, Visser AJ (2012) The impact of local food production on food miles, fossile energy use and greenhouse gas emissions: the case of the Dutch city of Almere. In: Viljoen A, Wiskerke SC (eds) Sustainable food planning – evolving theory and practice. Wageningen Academinc Publishers, Wageningen, pp 115–126
Kempen M (2012) EU wide analysis of the common agricultural policy using spatially disaggregated data. Inaugural-Dissertation, Bonn. 96 pages
Kempen M, Heckelei T, Britz W, Leip A, Koeble R (2005) A statistical approach for spatial disaggregation of crop production in the EU. In: Arfini F (2005) "Modelling agricultural policies: state of the art and new challenges", proceedings of the 89th EAAE Seminar, pp 810–830
Kneafsey M, Venn L, Schmutz U, Balázs B, Trenchard L, Eyden-Wood T, Bos E, Sutton G, Blackett M (2013) Short food supply chains and local food systems in the EU. A state of play of their socio-economic characteristics. JRC scientific and policy reports. Rep. 25911, Joint Research Centre, Seville. 128 pages.
Litman T (2003) Transportation cost analysis: techniques, estimates and implications. VTPI. www.vtpi.org/tca
MacDonald E (2013) Pleasure drives and promenades: the history of Frederick Law Olmsted's Brooklyn Parkways. Columbia College Chicago Press. 240 pages
MRDH (2014) Agenda economisch Vestigingsklimaat. Concept-versie t.b.v. consultatie gemeenten. Metropoolregio Rotterdam Den Haag. http://mrdh.nl/sites/mrdh.nl/files/files/Agenda%20Economisch%20Vestigingsklimaat%20MRDH%20AB141219.pdf
Mücher CA, Klijn JA, Wascher DW, Schaminée JHJ (2010) A new European Landscape Classification (LANMAP) – a transparent, flexible and user-oriented methodology to distinguish landscape. Ecol Indic 10:87–103
Smeets PJAM (2009) Expeditie Agroparken. Ontwerpend onderzoek naar metropolitane landbouw en duurzame ontwikkeling. Ph. D. thesis Wageningen Universiteit, Wageningen 2009. – With ref. – With summary in English and Dutch. 346 pp
Smeets PJAM, Harms WB, van Mansfeld MJM, van Susteren AWC, van Steekelenburg MGN (2004) Metropolitan delta landscapes. In: Tress G, Tress B, Harms WB, Smeets PJAM, van der Valk AJJ (eds) Planning metropolitan landscapes; concepts, demands, approaches. Alterra Green World Research, Wageningen, pp 103–113: [s.n.]
Steen-Olsen K, Weinzettel J, Cranston G, Ertug Ercin A, Hertwich EG (2012) Carbon, land, and water footprint accounts for the European Union: consumption, production, and displacements through international trade. Environmental science and technology. ACS Publication, Report 10833. 9 pages
Taylor PJ (2012) Transition towns and world cities: towards green networks of cities. Local Environ 17(4):495–508
Terluin IJ, Kamphuis BM, Oudendag DA, van Leeuwen MGA (2013) Voedselvoorziening in Nederland onder buitengewone crisisomstandigheden. LEI-Rapport 2013–012. Den Haag. 100 pages
Tress G, Tress B, Harms WB, Smeets PJAM, van der Valk AJJ (eds) (2004) Planning metropolitan landscapes; concepts, demands, approaches. Alterra Green World Research, Wageningen, pp 103–113
van Eupen M, Metzger MJ, Pérez-Soba M, Verburg PH, van Doorn A, Bunce RGH (2012) A rural typology for strategic European policies. Land Use Policy 29(3):473–482
van Gelder JW, Herder A (2012) Soja Barometer 2012. Een onderzoeksrapport voor de Nederlandse sojacoalitie. Profundo, Amsterdam. 83 pages
van Latesteijn HC (2008) TransForum: organizing the transition towards metropolitan agriculture. Executive summary of a presentation at the agriculture in transition conference. Wageningen 26–28.10.2008
van Steekelenburg M (2009) Metropolitan agriculture. Plan van Aanpak. Concept. Transforum. Xplorelab. Zoutemeer. 16 pages

Verzandvoort SE, Arets E, Hack-ten Broeke M (2013) Werken aan een handreiking voor het berekenen van een ecologische voetafdruk Eindrapport KBIV WOt 2013, Alterra, Wageningen UR, Alterra Report 2554. 92 pages

von Thünen JH (1826) The Isolated State. English translation by Carla M. Wartenberg, with an introduction by the editor. Pergamon Press. 1966

Wackernagel M, Rees W (1996) Our ecological footprint: reducing human impact on earth. New Society Publishers, Gabriola Island

Wascher DM (2012) SUSMETRO – ein Steuerungsinstrumentarium zur Erhöhung der Versorgungseffizienz metropolitaner Landschaften als Beitrag zur Nachhaltigkeit städtischer Ballungsräume. In: Heiland S, Demuth B (eds) Tagungsband der Veranstaltung, Landschaft 2030′ des Bundesamtes für Naturschutz. Vilm in Zusammenarbeit mit der Technischen Universität, Berlin

Wascher DM, Roos-Klein Lankhorst J, Agricola H, de Jong A (2010) Impact assessment tools for food planning in metropolitan regions. Final report from the SUSMETRO Project, TransForum, Zoetermeer. 68 pages

Wascher DM, Zasada I, Jeurissen L, Arciniegas G, de Kroes J, Sali G, Corsi S, Monaco F, Schmutz U, Glavan M, Pohle D, van Eupen M (2015) Indicators, tools and method for metropolitan footprint tools. Deliverable report D3.3, FoodMetres, FP7 ResearchProject, 57 pages (not published)

Chapter 7
Food Cells and Food Nodes. Two New Concepts for Rethinking Traditional Urban and Food Planning Practices. The Case of Barcelona's Metropolitan Region

Sonia Callau, Josep Montasell, and Andreu Vila

Abstract The intersection between urban planning and food has aroused a growing interest in the past few years in the areas of research and policy. Extensive literature is devoted to food systems, alternative food networks and food planning. Much of this literature emphasises consideration of food on a local scale and the importance of using food to reconnect the city with its agricultural periphery. The main arguments in favour of this new paradigm are food security and food self-sufficiency and sustainable and resilient development of cities. However, food production in the rural-urban fringe is still threatened, and calls into question the sustainability per se of local systems. A rapidly emerging vision with significant support amongst professional planners and local communities is the integration of food issues in urban planning. This paper proposes a new methodology for urban planning in relation to cities' nearest agricultural area (*food cells*) and *agro-urban planning systems* based on calculating *food provision* and optimization of the flow of food between the two systems. The theoretical model is tested in the city of Barcelona. The conclusion is drawn that sustainable food systems are only possible if urban planning incorporates the food vector and only if done so from a systemic interpretation of the rural-urban relationship.

Keywords Food provision · Agro-urban system · Food planning · Food system · Ecological urbanism · Peri-urban agriculture

---

S. Callau (✉) · J. Montasell
Agroterritori Foundation, Girona, Spain
e-mail: scallau@agroterritori.org; jmontasell@agroterritori.org

A. Vila
Universitat Politècnica de Catalunya, Barcelona, Spain
e-mail: andreu.vila-pascual@upc.edu

C.-T. Soulard et al. (eds.), *Toward Sustainable Relations Between Agriculture and the City*, Urban Agriculture, https://doi.org/10.1007/978-3-319-71037-2_7

7.1 Introduction

Agricultural areas and food have historically been an essential part of urban and, over time, spatial planning. Planners such as Cerdà (1867) and Howard (1902) were promoters of city models that incorporated rurality or food within the urban space. It's what Cerdà called the 'ruralisation of cities' (Soria y Puig 1999), while Howard proposed the 'garden city'. However, the growth of cities produces a paradoxical effect (Jarosz 2007) in the urban-rural relationship and its planning. As a city grows, it occupies the surrounding agricultural area and stops producing the food supplied to it. Consequently, a city ceases to have the capacity to produce its own (food) resources, which must then be imported from other production areas. Food is gradually becoming less and less associated with nearby agricultural land and something invisible (Steel 2008), at least in the way cities are configured and, therefore, in the way they are planned (Pothukuchi and Kaufman 2000; Sonnino 2009). At the same time, the food system is changed, as are the organisation of urban and agrarian structures. Farms grow in size and specialization (European Commission 2013) to be able to compete in the global market; small businesses are progressively replaced by large shopping centres, primarily in Northern Europe (Kneafsey 2015); and the distance food travels increases, as does that of the consumer to the point of sale. This Chapter will consider why –urban- food planning is so important and, in doing so, will articulate a new proposal to build more sustainable and resilient food systems.

There is a considerable body of literature dedicated to food systems and food planning (among others, Pothukuchi and Kaufman 2000; Born and Purcell 2006; Weber and Matthews 2008; Poirier and Théveniaut 2008; Donovan et al. 2011; Marsden and Sonnino 2012; Granvik 2012; Morganti 2011). For decades, food systems have been defined as *the chain of activities beginning with the production of food and moving on to include the processing, distributing, wholesaling, retailing and consumption of food and eventually the disposal of waste.* For the purpose of this study, we define *local food systems* as those characterised by three domains of *proximity* or *locality*: geographical proximity, relational proximity and values proximity (Eriksen 2013) between the area where food is produced and the area where this food is consumed.

This Chapter discusses *local food systems* as a strategic policy in metropolitan terms, not only because of their environmental or food security assets, but for their contribution to more 0072esilient rural-urban systems. Strategies adopted in Barcelona (Catalonia, Spain) to apply the precepts of *ecological urbanism* in what is called the *superblocks* are considered and used as a basis on which to develop a theoretical model of food planning, based on the design of *urban* and *food cells* and their interconnection. Why Barcelona? Barcelona and its peri-urban agricultural space, serves as a laboratory for implementing the proposed theoretical model and for extracting implementation measures that could be extrapolated to other cities and be used by planners and policy-makers. There are two features that make Barcelona particularly recommendable as a testing ground. On the one hand, it is the only European city with five superblock projects (equivalent to the idea of eco-neighbourhoods) in progress and one that is operating as such (Gràcia) and, secondly, it is considered as one of the few European cities where peri-urban agricultural areas have been preserved as vibrant agrarian land and claimed by a part of civil society as a food resource for the city (Paül and McKenzie 2013). The Gràcia superblock and the horticultural space in the Baix Llobregat area, the closest one to this urban space and located just 11,4 km south of it, are the areas to which the proposed theoretical model is applied. Although this study is limited to a space exclusively for production of fruits and vegetables, it is on target because they are the agricultural products most directly related to *Alternative Food Networks* (AFNs) and cities' agricultural belts, while also being the most difficult to plan due to their perishable nature.

This Chapter is drawn on documentary analysis undertaken by the authors in metropolitan Barcelona regarding urban and spatial planning, peri-urban agriculture, local food systems and food consumption patterns. Additional data were generated from in-depth interviews with 43 local farmers in 2009–2010 regarding production and commercialization systems of the farms and field observations collected by the authors between 2005 and 2010 to obtain production data (kg/ha) of various fruit and vegetable crops in the Baix Llobregat agricultural area.

This Chapter explains the characteristics of *ecological urbanism* and its weaknesses in relation to food systems sustainability and long-term resilience. It then turns to the case study of Barcelona, with particular attention paid to the Gracia District. The following section develops the theoretical model, which proposes a systemic approach to food planning that takes into account social, economic, cultural and environmental concerns and bridges the gap between food, urban and spatial planning interconnections. This theoretical framework is founded on the precepts of *ecological urbanism* and proposed planning of *agro-urban systems* which incorporate the food vector. The Chapter ends with a critical discussion of the theoretical model and a proposal for its implementation and concludes by arguing that urban planning could play a fundamental role for integrating agriculture and food into resilient and sustainable cities and metropolitan areas.

7.2 Ecological Urbanism and Its Contribution to the City-Food Relationship: A Theoretical Framework

As noted by Mostafavi and Doherty (2010), cities are spaces of complex processes (economic, political, social and cultural) and ecological urbanism is the tool able to solve conflicts inherent to the city/ecology relationship. Ecological urbanism, beyond successfully synthesising the conceptual corpus of the discipline itself (compactness, mixed-use, sustainable mobility and management of metabolic fluxes), incorporates the systemic vision of urban spaces, characterised by analysing the city as a system consisting of a set of units that interact (Gaffron et al. 2005). Research in urban ecology increasingly considers the consumption of external resources (water, energy, land), planning of urban and regional structures, transportation and the flows of energy and materials (Hernández et al. 2009). In fact, when reference is made to ecological urbanism mention must be made of regulating a city's own metabolism: cutting inputs of energy and materials by using local resources, cutting outputs in the form of waste and developing an organizational strategy of urban systems based on information and knowledge. According to this model, how cities are formed is not limited to the urban dimension, but also to their reliance on the nearby territory (Terradas 2001).

In the Barcelona context, Rueda et al. (2012) have developed and applied the precepts of ecological urbanism to the city of Barcelona in what they call the *superblocks*, a contemporary adaptation of Cerdà's model of *blocks* (100 m × 100 m). One of the characteristics that make it particularly interesting is its proposal for urban regeneration. Namely, Rueda incorporates the traditional criteria of ecological urbanism to already built spaces, in this case, the Cerdà block, sized to allow getting around on foot and to make an accessible quality urban space compatible with a road network with traffic. Under this premise, a new elementary *cell* of 400 m × 400 m is proposed, which is called *superblock*, defined as the 'basic cell' for structuring a city. Rueda very graphically defines the idea of superblocks as a "small town with its own character". The concatenation of superblocks creates a network of basic roadways where motorised traffic passes, thereby freeing up the space inside, with the aim of public space being free of traffic (Rueda et al. 2012). These superblocks become new city centres, which increase public space, the mix of uses and functions and people's accessibility to various services.

Beyond the organisational aspects, Rueda introduces two strictly ecological dimensions: *biodiversity* and *urban metabolism*. Biodiversity, linked to urban green spaces, is defined as a minimum amount of green space per inhabitant (10 m^2/inhabitant) and works via a network of green spaces accessible in an everyday manner. Urban metabolism, however, seeks maximum self-sufficiency from 'local metabolic resources', which include *fostering sustainable production systems of staple foods on a local scale and more efficient and healthy consumption patterns* (Rueda et al. 2012). The city and rural area are analysed as an 'agro-urban system'. This vision goes beyond seeing individual components of the food system (Sonnino et al. 2014), and aims to seek new organisational and relational formulas for agro-urban systems (from production to consumption), where food is the connection vector. This new perspective highlights the need for comprehensive and cooperative planning beyond city limits and one that involves the entire urban system. In this regard, it can be stated that although it is true that superblocks have been successful models in integrating environmental and social criteria, their transfer onto a territorial scale is still a major challenge. The main idea developed in this research is how food could be integrated into the proposal of superblocks.

7.3 Case-Study Area

7.3.1 Barcelona's Food System: A General Overview

The Barcelona Metropolitan Region (BMR) still has an important agricultural area (Paül and Tonts 2005) in spite of the significant reduction in recent decades due to urban growth (Paül and McKenzie 2010). In the year 2009, nearly 20% of the land was agricultural, equivalent in absolute values to 50,000 ha (Institut d'Estadística de Catalunya 2009). One of the most distinctive features of the metropolitan agricultural area is the wide range of agricultural and livestock production: horticultural production in the coastal plains, cereals in the interior plains, vineyards on the southern edge and fruit trees on the plains and mountains south of the city.

The BMR has a population of about five million inhabitants (Institut d'Estadística de Catalunya 2014), of which nearly two million (40% of the metropolitan population) live in the city of Barcelona. In the past decade, the population has increased substantially, and agriculture has declined: from 1989 to 2009, the BMR has lost 9719 ha of farmland and 30% of farms.

The region is characterised by significant production of vegetables with high demand from local markets. The horticultural production belt around Barcelona (the Baix Llobregat and Maresme counties) accounts for 31.24% of the total quantity of vegetables produced in all of Catalonia (Institut d'Estadística de Catalunya 2009). It is estimated that around 75% of what is grown in the Baix Llobregat area is allocated to metropolitan consumption (Paül and McKenzie 2013). Barcelona's large international fruit and vegetables wholesale market (Mercabarna) was established in 1971. It is a central node for import and export of food from around the world, but it is also a point of sale for products from throughout Catalonia (15.5% of fruit and vegetables marketed at Mercabarna originate in Catalonia and 65.9% in Spain).

Food sector establishments such as restaurants, local markets, grocery stores, supermarkets and specialty food stores are many and significant in the city of Barcelona. The city of Barcelona has devoted numerous efforts to consolidate local markets within the city and to protect small food grocery stores against big malls and supermarkets. In the city alone, 38 fixed food markets are distributed throughout every district in the city. In general, it can be said that every inhabitant can reach a municipal market on foot in no more than 15 min. There is an average of one market per 42,656 inhabitants; 41.7% of products sold in local markets come from Mercabarna, whereas only 8.3% come from local farmers for direct purchase. 85.8% of buyers who go to municipal markets do so on foot, and 5.4% go by public transport. The average time spent in travelling to markets is 9.3 min (Ajuntament de Barcelona 2009).

In addition to a mounting interest in local and short supply chains, there are an increasing number of food buyers focusing on expanding local food purchasing. Through an organization called *Associació de menjadors escolars ecològics* (ecological school canteen association), many schools in Barcelona have been expanding their connections with BMR farmers. There has been, as well, exceptional growth in the number of consumers' cooperatives during the last 20 years. Today, 50 consumers' cooperatives provide food and other products to their members in a self-managing and agro-ecological manner. Each has, on average, a total of 32 members (families) who represent, in the entire city, a total of 1600 members affecting 4500–5000 people. Average weekly spending is about €24, representing a gross income of €41,472/cooperative/year.

Taking the average per person/per year consumption of vegetables, fruit and potatoes, Barcelona city annually consumes 120,000 tons of vegetables and some 180,000 tons of fresh fruit. The sum of these two amounts represents 34% of total vegetables, fruit and fresh potatoes sold annually by Mercabarna.

7.3.2 The Gràcia Superblock in the Context of Barcelona

The Gràcia superblock is located in the *Vila de Gràcia* (Fig. 7.1), covers an urban area of 22.56 ha (475 m × 475 m) and has an estimated population of 9000 inhabitants. The Gràcia superblock is characterised by an urban grid form of narrow streets, buildings of relatively low height (ground floor plus two or three floors), and few green spaces (Fig. 7.2). There are numerous public squares that act as

Fig. 7.1 Situation of the Gràcia superblock (yellow square) in relation to the Vila de Gràcia (yellow area) and the city of Barcelona. © BCNecologia. Reproduced with permission

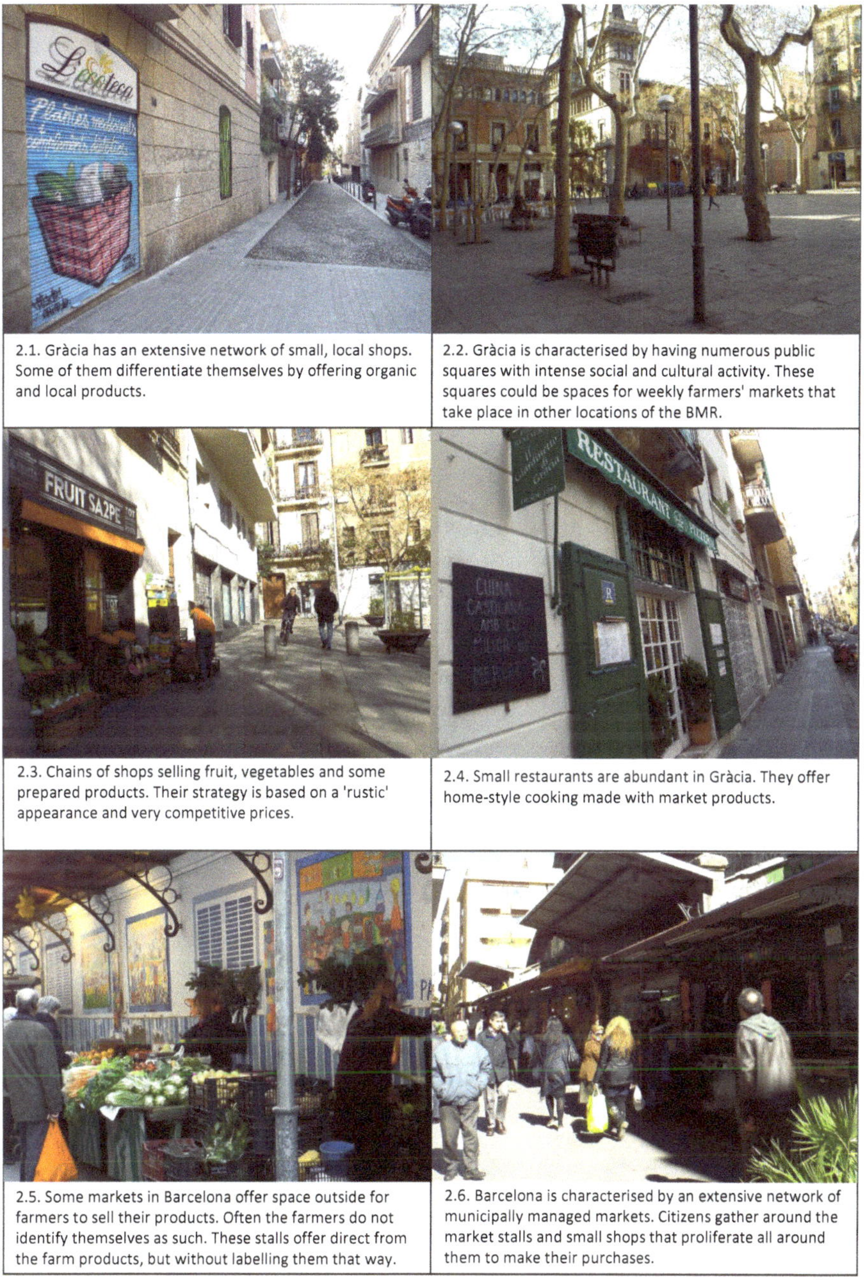

2.1. Gràcia has an extensive network of small, local shops. Some of them differentiate themselves by offering organic and local products.

2.2. Gràcia is characterised by having numerous public squares with intense social and cultural activity. These squares could be spaces for weekly farmers' markets that take place in other locations of the BMR.

2.3. Chains of shops selling fruit, vegetables and some prepared products. Their strategy is based on a 'rustic' appearance and very competitive prices.

2.4. Small restaurants are abundant in Gràcia. They offer home-style cooking made with market products.

2.5. Some markets in Barcelona offer space outside for farmers to sell their products. Often the farmers do not identify themselves as such. These stalls offer direct from the farm products, but without labelling them that way.

2.6. Barcelona is characterised by an extensive network of municipally managed markets. Citizens gather around the market stalls and small shops that proliferate all around them to make their purchases.

Fig. 7.2 Photos of the streets of the Gràcia superblock. 2.1 (Pictures by the Author 2015-03-24)

inter-relationship spaces, and which replace to some extent the lack of open spaces. The objectives for implementing the Gràcia superblock model, and those related to the purpose of this study would be:

Sustainable mobility, by freeing up public space initially allocated to motorised vehicles (62% pavement and 38% of carriageways) in favour of mobility by foot traffic and cycling (95% of pavement and 5% of carriageways). Only residents are allowed to use their vehicles within the superblock; in priority areas for pedestrians, traffic speed is limited to 10 km/h.

Design of *urban concentration systems* (UCS) in order to concentrate the bulk of transit of large, heavy vehicles into specific strategic points of the urban network. Capillary distribution of short food supply chains from the UCS to outlets would be done with small size vehicles with low environmental impact. The size and location of the UCSs would depend on the number of economic activities per linear metre and the number of weekly operations of each economic activity.

Accessibility to fresh food stores (fruits and vegetables), with distribution throughout the superblock that guarantees access on foot in no more than 7 min.

Gràcia has an extensive network of small, local shops. Some of them differentiate themselves by offering organic and local products. 2.2. Gràcia is characterised by having numerous public squares with intense social and cultural activity. These squares could be spaces for weekly farmers' markets that take place in other locations of the BMR. 2.3. Chains of shops selling fruit, vegetables and some prepared products. Their strategy is based on a 'rustic' appearance and very competitive prices. 2.4. Small restaurants are abundant in Gràcia. They offer home-style cooking made with market products. 2.5. Some markets in Barcelona offer space outside for farmers to sell their products. Often the farmers do not identify themselves as such. These stalls offer direct from the farm products, but without labelling them that way. 2.6. Barcelona is characterised by an extensive network of municipally managed markets. Citizens gather around the market stalls and small shops that proliferate all around them to make their purchases.

7.4 A New Theoretical Tool to Incorporate the Food Vector in a City's Metabolism

This section discusses a new theoretical formulation of the superblocks concept proposed by Rueda et al. (2012), which includes the *food vector*. In this regard, the proposed model, i.e. the agro-urban system adds to the restructuring of urban space, the restructuring of the nearby agricultural space and flows between both of them. It aims to create a new tool that, from an urban planning point of view, enables incorporation of the food vector in a city's metabolism. The functionality of this metabolism is based on the association of each urban cell with a food cell or food super-parcel to ensure a city's inhabitant with a food supply. Food cells can be understood as an 'infrastructure' (Fleury and Moustier 1999), 'structure of food supply' (Getz 1991) or 'municipal food terminal' (Pothukuchi and Kaufman 1999) for cities. Urban cells and food cells are conceived as subunits of the same agro-urban system, and understood as the "reconnection" of the city with its nearest productive area. In short, an avant-garde rural-urban relationship that aims to integrate productive space, urban-consumer space and how they relate in the same agro-urban system.

Before advancing to the theoretical formulation, the essential components of agro-urban systems are briefly described:

Food cell refers to the food production system (farmland and farmers) required to produce the food needed to supply a given superblock or urban cell.

Urban cell refers to the portion of the urban system on which agricultural space requirements needed to supply the population are calculated. In terms of urban planning, each urban cell incorporates a minimum module of agricultural land or *food provision area* (food cell) in order to meet its food needs.

Food Node corresponds to the strategic point(s) of superblocks with the best space opportunity (in relation to the production and consumption space) for concentration of food. In terms of urban planning, node design seeks to minimise distances and intensities of food shipment (from production space to the point of retail sale), loading and unloading and public accessibility to points of sale. The distribution of nodes must ensure coverage of food needs of an urban cell. Moreover, food nodes become essential for application of modern logistics practices tailored to the needs of the ecological urbanism.

The *agro-urban system* corresponds to the agricultural area (food cell), urban areas (urban cell) and their joining via food (food node). The agro-urban system analyses the elements and fluxes that make up the food system. It is based on the optimisation of: production space (agrarian production, soil protection, and a model of farmers and farms); the concentration of food at delivery destinations (dimension and logistics); the co-distribution from the production space to the logistics platform or UCS in the superblocks (vehicle type, quantity shipped, distance and total time); distribution from the logistics platform to the point of sale; consumer mobility from the point of consumption to the point of sale and, as a result, exchange and communication.

7.4.1 *Food Cells: Calculation and Design*

To calculate the size of the food cell, data is used in relation to: population density of the superblock or urban cell (inhabitants/urban cell), average annual consumption of fruit and vegetables (kg/urban cell), and production of fruit and vegetables (kg/m^2) in the production area with which a specific urban cell is linked. The climatic and edaphologic features of the agricultural land are taken into account when calculating production. Production data relate to farms managed agro-ecologically, based on two fundamental criteria: diversified production of fruit and vegetables by means of mixing crops to increase biodiversity (Raigón et al. 2006, Ninoska and Onesias 2011) and a holistic vision of the agricultural area, understood as a closed system, where crop rotation, reservoirs of auxiliary wildlife and the incorporation of organic matter are key to ensuring its stability. These data could be extrapolated to a production area located within the Mediterranean basin, with fertile soils, and a type of production that meets these characteristics: ability to cultivate two cycles of vegetables per year (conservative approach in that the climate makes possible three or even sometimes four cycles), estimating a production of 25 tons/ha per cycle per year (a total of 2 cycles, and 50 tons) of vegetables production; production of 20 tons/ha of fruit annually; establishment of mixed crops (combined fruit and vegetable production in the same parcel), with production per hectare of 51 tons of fruit and vegetables (Vila 2010).

Based on these data, the 'per person annual food provision' corresponds to a surface area for growing vegetables of 24 m^2/inhabitant/year and 75 m^2/inhabitant/year of fruit. With a polyculture or mixed culture, surface needs are reduced to around 40% and are placed at 53 m^2/inhabitant/year. One hectare of mixed farming (fruit and vegetables together) is capable of supplying fruit and vegetables for 189 people, whereas for fruit and vegetable crops grown separately this figure would be reduced to 101 people. The 53 m^2/inhabitant/year needed would have to be increased by 30%, attributable to the space that is allocated to agricultural services (irrigation facilities, roads, margins) that cannot be counted as a productive land. This value is obtained from extensive fieldwork undertaken in 2007–2008 about agricultural land-uses of the Baix Llobregat Agricultural Park (Paül 2008). According to the data obtained, two thirds of the land was dedicated to agriculture while the rest (1/3) corresponded to non-cultivatable area (irrigation canals, roads, agricultural buildings or non-agricultural uses). The resulting value would be 70 m^2/inhabitant/year (Table 7.1).

Table 7.1 Data for calculating needs for fruit and vegetables per inhabitant and year

Crop	Production per hectare (kg)[a]	Consumption person/year (kg/year)[b]	Inhabitants that can be fed from 1 ha	Surface area needed per inhabitant/year (m^2/inhab/year)
Vegetables	50,000	120	417	24
Fruits	20,000	150	133	75
Fruit and vegetables (separately)	–	–	101	99
Mixed crops	51,000	270	189	53
Mixed crops + correction rate (+30%)				**70**

[a]Field data collected between 2005–2010 based on observations and notes aimed at obtaining data on production yields (kg/ha) of various fruit and vegetable crops grown following agro-ecological production criteria, and mixed vegetable and fruit crops in the Baix Llobregat area

[b]Annual per capita consumption per year of fruits and vegetables are based on 2010 consumption data for Catalonia. These amounts have been increased by 25% (fruits) and 20% (vegetables) to include tinned food consumption

The food provision values (for fruit and vegetables) vary depending on the data on consumption/inhabitant/year, and according to production estimates (either due to the management system, or the geographic area where they are located). Values provided by other sources, range from 60 m^2/inhabitant/year (World Health Organization and FAO 2005), to 60–100 m^2/inhabitant/year (proposed by the *Association pour le maintien de l'agriculture paysanne* or AMAP) up to 470 m^2/inhabitant/year (Kastner et al. 2012).

7.4.2 Food Nodes

In the current debate on shipping and distribution of local food to the end consumer, food hubs are presented as one of the solutions to facilitate a greater connection between producers and consumers (Matson and Thayer 2013) and also to expand the scope of AFNs beyond direct sales, by offering services that allow them to compete with large-scale distribution while maintaining the identity of producers (Le Blanc et al. 2014). The model we are proposing includes the concentration of food in urban food platforms (food nodes), which, unlike food hubs, are characterised by their 'proximity' to the point of sale, because they are located within the urban space. Thus, in our model, each superblock has a food node sized according to the density and type of final food distribution points.

The food node concept has been developed as a way to connect different producers and different types of foods with 'buyers' (restaurants, retail stores, municipal markets, school canteens and catering in general) or 'consumer groups' (consumer cooperatives, consumer groups) in a more efficient and sustainable form. The functional scheme for a platform includes, beyond the definition of management formulas (private, public or public-private), the breakdown of the entire operational structure: reception, registration and classification of food, storage, organisation of deliveries and distribution to the final points of distribution. This structure is adapted in each case to the specificity of each superblock, and to catch each type of buyer and consumer groups belonging to a food node.

7.4.3 Distribution of Fruit and Vegetables in the Gràcia Superblock

We will now analyse the specificities of the Gràcia superblock in relation to the type and distribution of points of sale of fruits and vegetables and the importance of local produce in its food system. A careful analysis of the Gràcia superblock shows high diversity of systems for selling fruit and vegetables, and an increase in AFN. In terms of accessibility to greengrocers, the data show a high potential for implementation of short food supply chains, which is not the case in other districts in Barcelona

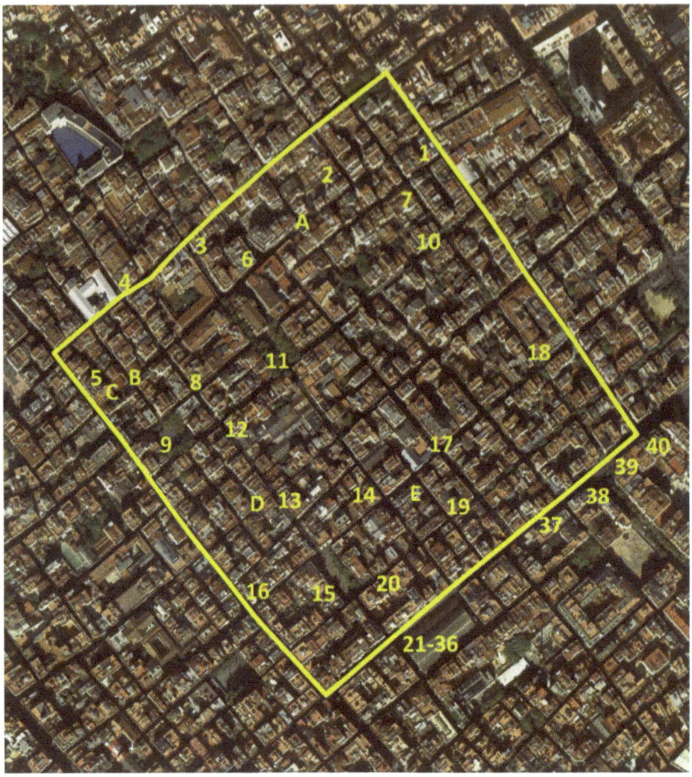

Fig. 7.3 Distribution of points of sale of fruits and vegetables in the Gràcia superblock. Each number represents a point of sale. The letters represent consumer cooperatives

(see Fig. 7.3). There are a total of 12 shops selling mainly fruits and vegetables, 11 shops selling food stuffs, and 15 stalls selling fruits and vegetables within the local market and 5 consumer co-ops (and three more in the vicinity).

The walking distance to the food selling points is about 120 m and the biggest distance is 297 m. It is estimated that current sales points may supply 16,000 people with fruit and vegetables, a figure that exceeds the current number of residents. Based on this result, it can be stated that the number of points of sale of fruit and vegetables in the Gràcia superblock is sufficient to allow citizens to reach these outlets without having to travel outside of it. In regard to food distribution to the point of sale, it should be emphasised that there is no type of planning or coordination. Nearly 50% of products sold in the municipal market are supplied by Mercabarna (Ajuntament de Barcelona 2009). Municipal market vendors travel individually to Mercabarna to purchase the fruit and vegetables they sell at their stall. A similar model is used by supermarkets and shops. In the case of consumer cooperatives, it is the farmers who travel to the cooperative's premises.

With exception of the cases of consumers' cooperatives, it is difficult to quantify exactly what comes from local production out of the total of products marketed within the superblock. However, based on surveys of 43 farmers in the Baix Llobregat horticultural production area, it is known what distribution systems are used by today's farms and the potential to meet the demands of a superblock such as Gràcia. The results show that there is a wide variety of distribution systems and a predominance of direct sales (56.2% of companies) over sale through an intermediary (43.8%). It is also important to note that companies tend to combine different sales systems, and some that use direct sales also market part of their production via Mercabarna. While models combining direct sales and sales via Mercabarna allow farmers to manage surpluses, it should be stressed that the prices received (€/kg) for this sales channel are much lower than those obtained from direct sales.

7.5 Discussion: Food Cells and Food Nodes, a Theoretical Approach for a New Food Planning

The aim of this paper was to highlight the critical relationship between urban and food planning and propose a model for integrating agriculture and food into cities, in which urban planning can play a fundamental role. Our working hypothesis was that it is possible to incorporate the food vector into sustainable urban models, similar to eco-neighbourhood models or Rueda's superblocks. To verify our methodological proposal, we discuss the implementation of this model in the specific case of the Gracia Superblock. How would we connect the Gràcia superblock to its nearest agricultural area, fulfilling the premises of ecological urbanism? First, we would estimate the minimum size of the food cell we wish to connect to the superblock in question and, secondly, analyse the connection between the two. Based on the result of 'food provision' calculated in the previous section (70 m^2/inhabitant/year), an agricultural surface area of 63 ha and around 18 farmers (the average size of farms in the Baix Llobregat is 3.5 ha) would be required for a population of 9000 inhabitants in the Gràcia superblock. Every hectare of urban land of the Gràcia superblock needs nearly 3 ha of land in the Baix Llobregat to meet the needs of fruit and vegetables of its inhabitants (Fig. 7.4).

The calculation process to establish the amount of agricultural land needed demonstrated the importance of having accurate and detailed information on both the urban and the agricultural areas under study. Thus, small variations in the values of production yields or per capita food consumption give significantly different results. Likewise, it should be noted that small increases in production yields can significantly reduce the production surface required, following agroecological handling and management criteria. In this case, systems to support farmers could be developed to establish plot designs that, on the one hand, make it possible to optimize production yields, and on the other adjust products and harvest schedules to the demand of their market sector. Future partnerships between farmers could be implemented, for planning production and transportation, which currently represents a high cost in time and organizational difficulties.

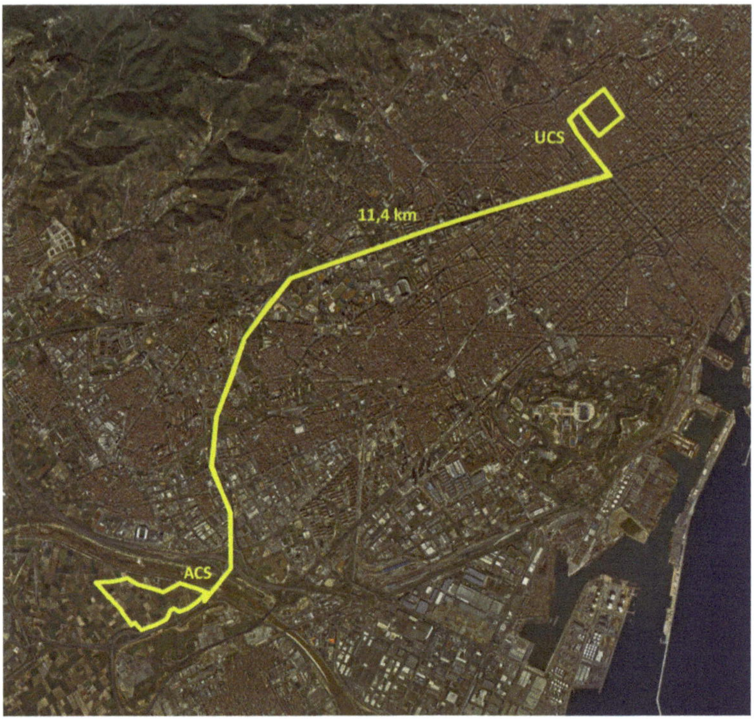

Fig. 7.4 Ratio of urban cells to food cells in the case of the Gràcia superblock. The areas marked in yellow represent the ratio of agricultural space required needed to feed the inhabitants of the Gràcia superblock. The shipping distance of food from the 'Agricultural concentration system' (ACS) to 'Urban concentration systems' (UCS) is 11.4 km (Source: Author 2015)

Based on the analysis of shipping and distribution of fruit and vegetables within and outside the superblock, significant deficiencies were detected in both systems for direct sale, as well as in other non-direct marketing channels. In the former (direct selling of food baskets to homes and consumer cooperatives), although there is planning in the concentration of food at the source, lack of concentration in their destination increases the distance travelled and, consequently, distribution time, loss of efficiency, increased costs for the company, a major environmental impact from emissions due to delivery and a negative contribution to the problems of traffic congestion in the city. In the particular case of the distribution of baskets to consumer cooperatives, the greater concentration of deliveries significantly minimises the impact of shipping. In the case of the flow of food moving through Mercabarna, it has been observed that many distributors and farmers are located nearby, travel similar routes (from their fields to Mercabarna) and serve the same customers (cooperatives or purchasing agents). The same situation occurs in the case of market stallholders or shopkeepers. Both cases repeat the previously mentioned impact from shipping and traffic congestion. All this highlights the unsustainability of current transportation and food distribution systems in the city, and raises major challenges in co-distribution and concentration. Food flow could improve with co-distribution systems from the production area to the UCSs proposed by Rueda. Multiple farmers would be coordinated to establish a common delivery route for fruit and vegetables to the superblock UCSs. Travel would only be required from the superblock's point of sales to the UCS.

It was also observed that, although the current distribution of points of sale for fruits and vegetables in the Gràcia superblock allows access on foot, most of 'local shops' are disconnected from the food production site and the farmers who produce it. In the future, awareness campaigns to consumers should be implemented having to do with the consumption of local products (as has been done regularly in recent years with local businesses) and the shops around the added value of providing them. Moreover, more information is needed about the difficulties of farmers for supplying their products to shops in the city and identifying them as 'local'.

In general, the results show a high potential for promoting local food systems in the Gràcia superblock. Support for partnership projects in the area of production and at source concentration and optimization of distribution to the point of sale are potential strategies for an economically efficient, environmentally sustainable and socially and culturally integrated food system. This case study serves as an example of how ecological urbanism can contribute to the 'relocalisation' of food systems and to reduce their environmental impact.

7.6 Conclusions

The model of superblocks proposed by Rueda and the theoretical proposal to integrate the food vector present opportunities for proactive development of sustainable urban food strategies which potentially will consolidate urban-rural relationships and contribute to city's food resilience. This research shows that the food vector cannot be integrated into ecological urbanism without a systemic vision of the rural-urban relationship. Food systems and urban systems in metropolitan contexts are highly complex, as are the difficulties in seeking effective formulas to reconnect the two. Our methodology for calculating the 'food provision area' is a first step for urban planners to incorporate in planning the need to preserve productive land to meet a city's food demand. The amount of agricultural land would not differ greatly from the 'provision of green spaces' that some European cities have. For example, in the case of Barcelona, there is provision for 10 m^2/person of green space, accessible from a distance of no more than 2 km. As shown in the case of the Gràcia superblock, the calculation is easy and very enlightening: each hectare of urban land should connect to 3 ha of agricultural land to obtain the amount of fruits and vegetables necessary to feed all its inhabitants. Once 'provision of agricultural land' is ensured, the proposal is to go one step further and plan the entire food system, from local food production until it reaches consumers. The flow of food could be managed in an innovative way, based on planning production and shipping of food from the closest agricultural area to the UCS, and from them to the selling areas.

This research highlights the potential of urban planning as a tool in helping develop local food systems through: analysis and improving of citizens' access to sales points of basic foods, UCS or food nodes design and establishment of provisions of agricultural land by means of protecting productive land needed to supply the citizens of its immediate urban environment. In addition, efficient logistics management would have a positive impact on farmers' income, in the increase in agro-diversity inherent in small and/or medium size farms, on the price of the product, on city's environmental quality and on the reduction of traffic intensity for shipment of goods. We should add that conserving cities agricultural belts is not just a food strategy, but also care of the territory and cultural identity.

This chapter provides a better understanding of local urban systems, and goes beyond partial approaches focusing on AFNs, food hubs or food. Further, the 'food nodes' proposal presents a wide range of opportunities for overcoming the discussion of the advantages and disadvantages of 'local' versus 'global'. Such urban food strategy would however require the various public areas of intervention (spatial planning, urban planning, health, education, commerce, economic development and agricultural production) partner with one another and with the private sector (agriculture, commerce, catering…).

This research has obviously limits (great difficulty in obtaining statistics on food production and consumption, origin of food consumed in the city, distribution of food produced locally) and the fact that it has only considered fruits and vegetables to test the model. Future studies should include all the foodstuffs produced within BMR (cereals, meat, poultry, milk and dairy products). Even though the model is based only in fruits and vegetables production, the theoretical approach proposed would potentially be transferable to other crops, using the same methodology and calculation.

The agro-urbanism proposal, which is based on knowledge, analysis and organization of local food systems in addition to being a theoretical proposal, also desires to be a framework for reflection on a model for cities autonomous in food resources, as well as being complex, diverse and rich. In this regard, food planning can become an effective tool in configuring a sustainable and resilient city model. The idea of reconnecting the 'urban cell' to a 'food cell' is not so much a physical fact but rather a conceptual one. The concept of the food cell adds rationality to town planning, making clear the need to always allocate a provision devoted to food production, and also the need to implement services to make possible the supply of produce by nearby agriculture. Despite being an invented 'model', it is based on the long experience in planning and management of the Baix Llobregat Agricultural Park and the realization that new food and territorial governance models are needed.

In short, the proposed model shows that it is possible to solve the limitations attributed to local food systems and incorporate the food vector into a compact, complex, organized and efficient city model.

References

Ajuntament de Barcelona (2009) Mercats de Barcelona. Activitats 2009. http://w110.bcn.cat/mercats/continguts/documents/memorialMMB202009.pdf. Accessed 10 Feb 2015

Born B, Purcell M (2006) Avoiding the local trap: scale and food systems in planning research. J Plan Educ Res 26(2):195–207. https://doi.org/10.1177/0739456X06291389

Cerdà I (1867) Teoría general de la urbanización y aplicación de sus principios y doctrinas a la reforma y ensanche de Barcelona. Imprenta Española, Madrid

Donovan J, Larsen K, McWhinnie J (2011) Food-sensitive planning and urban design: a conceptual framework for achieving a sustainable and healthy food system. Report commissioned by the National Heart Foundation of Australia (Victorian Division). http://www.heartfoundation.org.au/SiteCollectionDocuments/Food-sensitive-planning-urban-design-full-report.pdf. Accessed 29 Oct 2015

Eriksen SN (2013) Defining local food. Constructing a new taxonomy: three domains of proximity. Acta Agric Scand Sect B Soil Plant Sci 63:47–55. https://doi.org/10.1080/09064710.2013.789123

European Commission (2013) Structure and dynamics of EU farms: changes, trends and policy relevance. European Commission. EU Agricultural Economics Brief. http://ec.europa.eu/agriculture/sites/agriculture/files/rural-area-economics/briefs/pdf/09_en.pdf. Accessed 10 Feb 2015

Fleury A, Moustier P (1999) L'agriculture périurbaine, infrastructure de la ville durable. Cah Agric 8(4):281–287

Gaffron P, Huismans G, Skala F (2005) Ecocity: a better place to live. Facultas Verlags-und Buchhandels AG, Vienna

Getz A (1991) Urban foodsheds. Permaculture Activist 24:26–27

Granvik M (2012) The localization of food systems. an emerging issue for Swedish municipal authorities. Int Plan Stud 17(2):113–124. https://doi.org/10.1080/13563475.2012.672796

Hernández A, Velásquez I, Verdaguer C (2009) Ecobarrios para ciudades mejores. Ciudad y Territorio Estudios Territoriales XLI(161–162):543–558

Howard E (1902) Garden cities of tomorrow. Swan Sonnenschein & Co, London

Institut d'Estadística de Catalunya (2009) Cens agrari de l'any 2009. http://www.idescat.cat/pub/?id=censag. Accessed 30 Jan 2015

Institut d'Estadística de Catalunya (2014) Padró municipal d'habitants. http://www.idescat.cat/pub/?id=pmh. Accessed 30 Jan 2015

Jarosz L (2007) The city in the country: growing alternative food networks in Metropolitan areas. J Rural Stud 24(3):231–244. https://doi.org/10.1016/j.jrurstud.2007.10.002

Kastner T, Ibarrola MJ, Koch W, Nonhebel S (2012) Global changes in diets and the consequences for land requirements for food. PNAS 109(18):6868–6872. https://doi.org/10.1073/pnas.1117054109

Kneafsey M (2015) Discussion paper-focus group short food supply chain management. EIP-AGRI document. https://ec.europa.eu/eip/agriculture/en/content/discussion-paper-focus-group-short-food-supply-chain-management. Accessed 30 Sept 2015

LeBlanc J, Conner D, McRae G, Darby H (2014) Building resilience in nonprofit food hubs. J Agric Food Syst Community Dev 4(3):121–135. https://doi.org/10.5304/jafscd.2014.043.005

Marsden T, Sonnino R (2012) Human health and wellbeing and the sustainability of urban-regional food systems. Curr Opin Environ Sustain 4:427–430. https://doi.org/10.1016/j.cosust.2012.09.004

Matson J, Thayer J (2013) The role of food hubs in food supply. J Agric Food Syst Community Dev 3(4):43–47. https://doi.org/10.5304/jafscd.2013.034.004

Morganti E (2011) Urban food planning, city logistics and sustainability: the role of the wholesale produce market. The cases of Parma and Bologna food hubs. Dissertation thesis, Università di Bologna

Mostafavi M, Doherty G (2010) Ecological urbanism. Lars Müller Publishers, Baden

Ninoska S, Onesias D (2011) Un sistema de siete hortalizas en policultivos con cuatro dosis de fertilización evaluando rendimientos, uso equivalente del terreno (UET), UET financiero y de nutrientes. Proyecto especial de graduación. Escuela Agrícola Panamericana, Zamorano, Honduras

Paül V (2008) Estudi de les explotacions agràries en l'àmbit del Parc Agrari del Baix Llobregat. Institut d'Estudis Territorials, Barcelona

Paül V, McKenzie FH (2010) Agricultural areas under metropolitan threats: lessons for Perth from Barcelona. In: Luck GW, Race D, Black R (eds) Demographic change in Australia's rural landscapes. Springer, Dordrecht, pp 125–152

Paül V, McKenzie FH (2013) Peri-urban farmland conservation and development of alternative food networks: insights from a case-study area in metropolitan Barcelona (Catalonia, Spain). Land Use Policy 30:94–105. https://doi.org/10.1016/j.landusepol.2012.02.009

Paül V, Tonts M (2005) Containing urban sprawl: trends in land use and spatial planning in the metropolitan region of Barcelona. J Environ Plan Manag 48(1):7–35. https://doi.org/10.1080/0964056042000308139

Poirier Y, Théveniaut M (2008) Boletín Internacional de Desarrollo Local Sostenible 48. http://developpementlocal.blogspot.com.es/2008_05_01_archive.html. Accessed 10 Feb 2015

Pothukuchi K, Kaufman JL (1999) Placing the food system on the urban agenda: the role of municipal institutions in food systems planning. Agric Hum Values 16:213–224. https://doi.org/10.1023/A:1007558805953

Pothukuchi K, Kaufman JL (2000) The food system. A stranger to the planning field. J Am Plan Assoc 66(2):112–124. https://doi.org/10.1080/01944360008976093

Raigón MD, García MD, Guerrero C, Esteve P, Domínguez-Gento A (2006) Influencia de la asociación de cultivos sobre la relación equivalente de suelo. In: Proceedings of the VII SEAE Congress held in Zaragoza, Spain, on September 18–23rd, 2008. http://www.agroecologia.net/recursos/publicaciones/publicaciones-online/2006/. Accessed 26 Oct 2015

Rueda S, De Cáceres R, Cuchí A, Brau L (2012) El Urbanismo ecológico: su aplicación en el diseño de un ecobarrio en Figueres. Agència d'Ecologia Urbana de Barcelona, Barcelona

Sonnino R (2009) Feeding the city: towards a new research and planning agenda. Int Plan Stud 14:425–435. https://doi.org/10.1080/13563471003642795

Sonnino R, Moragues A, Maggio A (2014) Sustainable food security: an emerging research and policy agenda. Int J Sociol Agric Food 21(1):173–188

Soria y Puig A (1999) Introducción a Cerdà. In: Soria y Puig A (Comp.) Cerdà. Las cinco bases de la teoría general de la urbanización. Electa, Barcelona, pp 23–49

Steel C (2008) Hungry city: how food shapes our lives. Chatto & Windus, London

Terradas J (2001) Ecologia urbana. Rubes, Barcelona

Vila A (2010) Assaig de camp de cultiu mixte al Baix Llobregat. Keynote address in the Baix Llobregat agricultural park event «Experiències pràctiques d'horticultura i fructicultura ecològiques» held in El Prat de Llobregat. December 2011

Weber C, Matthews HS (2008) Food miles and the relative climate impacts of food choices in the United States. Environ Sci Technol 42(10):3508–3513

World Health Organization & FAO (2005) Fruit and vegetables for health. Report of a joint FAO/WHO workshop on fruit and vegetables for health held in Kobe (Japan), 1–3 September 2004

Chapter 8
The Integration Between Mediterranean Cities and Agriculture in Local Projects. A Conceptual Framework

Salma Loudiyi, Christophe-Toussaint Soulard, Patricia Abrantes, Sylvie Lardon, Guilhem Mousselin, Pascale Scheromm, Elodie Valette, Ouassila Bendjaballah, and Françoise Jarrige

Abstract The return of agriculture as a subject of debate in the realm of urban sustainable development is a recent phenomenon. Its presence is anchored in the recognition of the multifunctional role of agriculture and its importance in the conception and success of a more sustainable city. This study proposes a conceptual framework to analyse the socio-spatial integration between cities and agriculture in Mediterranean countries. The concept of integration is seen as a temporal process that articulates the actors, the spaces, and the resources linked by a project. Using a

S. Loudiyi (✉) · S. Lardon
Université Clermont Auvergne, AgroParisTech, Inra, Irstea, VetAgro Sup, Territoires, Clermont-Ferrand, France
e-mail: salma.loudiyi@vetagro-sup.fr; sylvie.lardon@agroparistech.fr

C.-T. Soulard · P. Scheromm · F. Jarrige
Innovation, INRA, CIRAD, Montpellier Supagro, University of Montpellier, Montpellier, France
e-mail: christophe.soulard@inra.fr; pascale.scheromm@inra.fr; francoise.jarrige@supagro.fr

P. Abrantes
Universidade do Porto, CEGOT, Faculdade de Letras, Porto, Portugal
e-mail: patricia.crmabrantes@gmail.com

G. Mousselin
Bordeaux Montaigne University, UMR CNRS 5319 Passages, Pessac, France
e-mail: g.mousselin@gmail.com

E. Valette
ART-DEV, University of Montpellier, CIRAD, CNRS, University of Paul Valery, University of Perpignan, Montpellier, France
e-mail: elodie.valette@cirad.fr

O. Bendjaballah
Laboratoire RNAMS Oum El Bouaghi, University of Constantine 3, Constantine, Algeria
e-mail: ouassila.bendjaballah@univ-constantine3.dz

C.-T. Soulard et al. (eds.), *Toward Sustainable Relations Between Agriculture and the City*, Urban Agriculture, https://doi.org/10.1007/978-3-319-71037-2_8

collection of local projects that explore the interactive dynamics of urban and agricultural relationships, the authors propose a conceptual framework, which reveals four aspects of socio-spatial integration. The first aspect encompasses socio-spatial integration through agricultural practices within the city. The second aspect includes integration processes through a pivotal actor playing as an interface between agriculture and the city. The third aspect concerns the flows and mobilities of resources and actors between cities and agricultural countryside. The fourth aspect deals with planning processes where the construction of networks of actors, spaces and resources are significant. These categories highlight the multidimensional processes of integration and the differentiated effects. The analysis is based on case studies situated in France, Italy, Portugal, Algeria, and Morocco.

Keywords Urban agriculture · Integration · Local projects · Mediterranean cities

8.1 Introduction

The reconnection of cities and agriculture takes multiple forms, ranging from the inclusion of agriculture in urban planning, to the development of closer food supply networks, or to urban agricultural production. Understanding these diverse relationships between cities and agriculture, and their organization through a diversity of actors, is one of the keys to characterizing sustainable urban agriculture. This phenomenon is part of a profound transformation of the images and practices associated with urban and rural dynamics, their respective functions and the relationships between them. The construction of sustainable cities requires an open and inclusive perspective on agriculture, as much for its environmental, social, and health advantages as well as the economics associated with insuring the presence of a local food system, whether intra-urban or peri-urban. This integration of cities and agriculture implies reciprocal development between two systems that have previously been perceived as separate.

Several works describe the reconfiguration of links between cities and local agriculture, particularly in light of the implementation of land-use planning mechanisms (Jarrige et al. 2006; Loudiyi et al. 2011; Perrin et al. 2013) or urban agriculture projects (Poulot 2011, 2014; Giacché 2014). These studies show the transformations of agricultural spaces and activities in or near a city. The issues in Europe have changed over the course of a decade, evolving from the simple preservation of nature and the living environment (Kühn 2003) to a more complete integration of natural and agricultural space in the urban setting as a means of ensuring ecological consistency and continuity, or functions of social integration (Niwa 2009; Ernwein 2014; Scheromm et al. 2014; Tornaghi 2014). More recently, the arguments for the reintegration of local agriculture in the urban food supply chain have joined the dialogue (Viljoen and Wiskerke 2012; Paül and McKenzie 2013; Specht et al. 2014). By contrast, in some countries of the Maghreb, the environment is only gradually being included in public policies for urban planning, and the integration of agriculture in urban development is still marginal (Valette and Philifert 2014). Other studies examine the specific inclusion of agriculture in urban development (Viljoen 2005; De la Salle and Holland 2010; Viljoen and Bohn 2014), and study the operational modalities of the integration of agricultural spaces in the urban system and their contribution to human health and well-being, the environment, and the implementation of sustainable food systems. These works show that the reconnection of cities and agriculture is an on-going process, and achievements in integration are often the result of local projects at various scales.

In terms of urban development and agricultural concerns, these works often describe asymmetric relationships as a result of a city's perspective on the functions of local agriculture. They also reveal the processes of constructing the capacity for action among the various actors, whether in the realm of the civil society, or that of agriculture or in local governments. Although these studies begin to explain the transformation of image and practices and the way this affects the links between cities and agriculture, they have not clarified the socio-spatial processes that lead to the reconfiguration of these links. In this paper, we focus on a particular aspect of these relationships: integration projects. As a preliminary definition, an integrative project can be seen as an organized effort to engage urban and agricultural actors, both private and public, in a collaborative mechanism, which addresses urban and agricultural issues. In the framework of an international research project,[1] we have identified, analysed, and put into perspective, integration projects in five Mediterranean urban regions. The data collected offer an overview of projects in different urban and peri-urban contexts of southern Europe and the Maghreb. What do these projects reveal about the various forms of integration of cities and agriculture? What processes favour socio-spatial integration? Moreover, what are the effects on the relationships between cities and agriculture?

We have positioned this investigation specifically on the concept of integration seen as a temporal process that articulates the actors, the spaces, and the resources linked by a project.

8.2 Integration as a Socio-spatial Process: Theoretical Frameworks

Integration, as a word or a concept, is polysemic. Roger Lee (1993) defines integration as "the creation and maintenance of diverse and intense patterns of interaction and control between formerly more or less separate social spaces. Integration involves the bringing together of different systems of meaning and action founded in different sets of social relations (norms, means of communication, indicators of direction and value, structures of power, domination and subordination)". He adds,

[1]This research has been funded by the ANR (French National Research Agency) through the DAUME project (Sustainablility of Urban Agriculture in the Mediterranean, 2011–2015).

"Geography is central in this process. Integration takes place through – not merely across- space. [...] the creation of a space or spaces of integration is central to any form of integration; without such spaces, the social relations which constitute integration cannot themselves be constituted".

In order to understand how integration functions and what it actually represents, Ruiz-Tagle (2013) suggests a theory of socio-spatial integration. He assumes, as does Lee, that integration can have both positive effects such as the elimination of barriers and the opening of closed systems, and negative effects such as the loss of identity (during the process of assimilation) or a rupture in the integrity of a system. Thus, integration does not necessarily appear as a faultless process.

Noting the importance of the spatial dimension in the processes of integration, Ruiz-Tagle, proposes that socio-spatial integration manifests itself as a multi-dimensional relationship that can function independently and at different scales. He also identifies four fundamental dimensions: The first dimension is the spatial proximity between social groups. The second dimension concerns the functional relationships that allow access to opportunities (such as economic access to services). The third dimension encompasses the relationships between different social groups. These social groups maintain relationships defining power or influence, and may or may not be hierarchical (social networks, leadership, political participation). The final dimension involves the identification of a common ground, and constitutes an immaterial or symbolic dimension (such as an imaginary border, or landmark). These dimensions correspond in part to the theoretical framework of the economics of proximity (Gilly and Torre 2000): the first two dimensions can relate to geographic proximity, while the latter dimensions correspond to organizational or institutional proximity. While the theoretical framework of proximity economics focuses primarily on productive relationships (Torre and Zuindeau 2009), and emphasizes this productive dynamic, the theory of socio-spatial integration aims to uncover the mechanisms at work in the organization of space and the associated social dynamics. Socio-spatial integration can be seen as interactions, which tie together different dimensions, such as those identified by Ruiz-Tagle (2013), each having a set of spatial characteristics, and conditions that define the redistribution of resources and access to these resources.

8.3 Integration Through Local Agri-Urban Projects: Methodological Approach

Our interpretation of the links between cities and agriculture is based on the study of local projects that emphasize the intersection of agricultural and urban dynamics. We take a particular interest in the ways in which public and private actors participate in the sustainability of urban agricultural systems (Soulard 2014), and in the modalities of their governance. With an objective of generalizing the observed forms of integration, the analysis compares and contrasts Northern and Southern

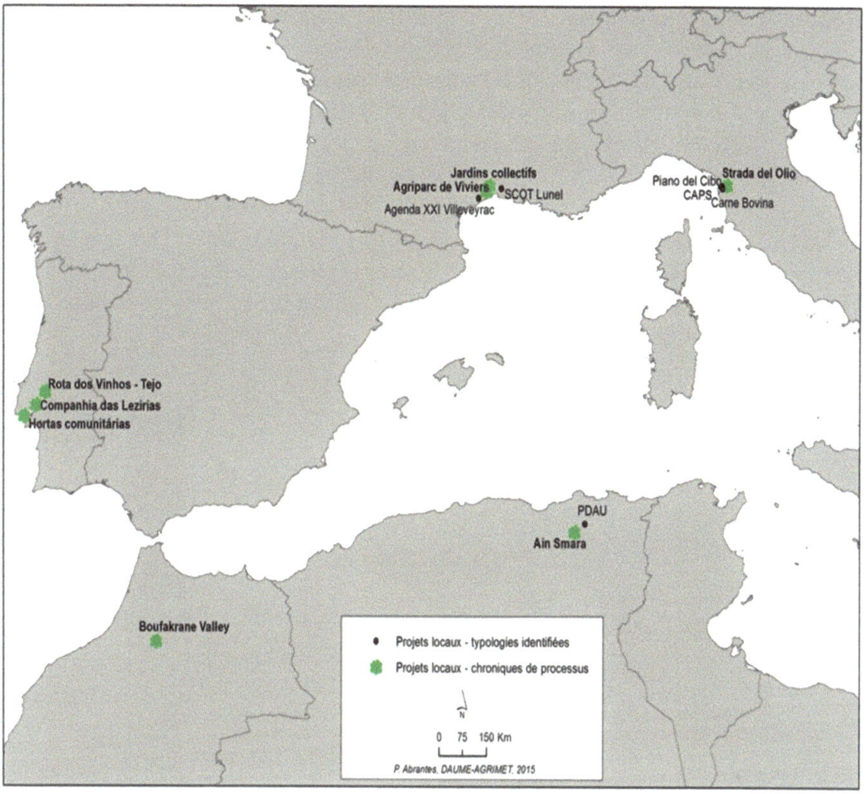

Fig. 8.1 Case studies locations

Mediterranean locations (Fig. 8.1). In our approach, we have concentrated on the analysis of local projects. This methodological choice, however, does not exclude other possible approaches (e.g. integration through practices and activities of larger agricultural exploitations).

The first stage consists of creating a classification of local projects. The characterization of cases selected is derived from a review of documents and interviews with local actors. We use this analysis to identify sustainability issues, which motivate these actors and translate, more or less effectively, into various mechanisms of action (e.g. Marraccini et al. 2013; or Perrin et al. 2013). Undertaken on each terrain selected for study, these analyses produce a collection of 14 local projects identified as being capable of responding to the issues surrounding the sustainability of urban agricultural systems, either in a formal or informal manner, intentionally or unintentionally. This initial characterization, however, gives static results, which do not contribute to the understanding of the temporal evolution of the projects.

The second stage seeks a more thorough analysis of certain projects in order to understand their functioning and governance. We select a number of local projects, which illustrate the diversity of processes at work, and provide an opportunity to perform an in-depth analysis. To accomplish this, we rely on method that reconstructs each case following a timeline that reveals the successive phases of the project and the associated contexts (Paoli and Soulard 2003): using semi-direct interviews with actors from each project,[2] and complementing them with document research, we characterize the project's emergence, life cycle, principal actors, achievements, and modalities of participation, as well as the controversies, which are woven into and alter the project.[3]

The information collected during the first stage is represented in a grid of two axes: actors and terrains (Fig. 8.2). Each project is placed on this grid as a function of the supporting actors in their degree of influence on the project, and the terrains in their position along the urban/rural gradient.

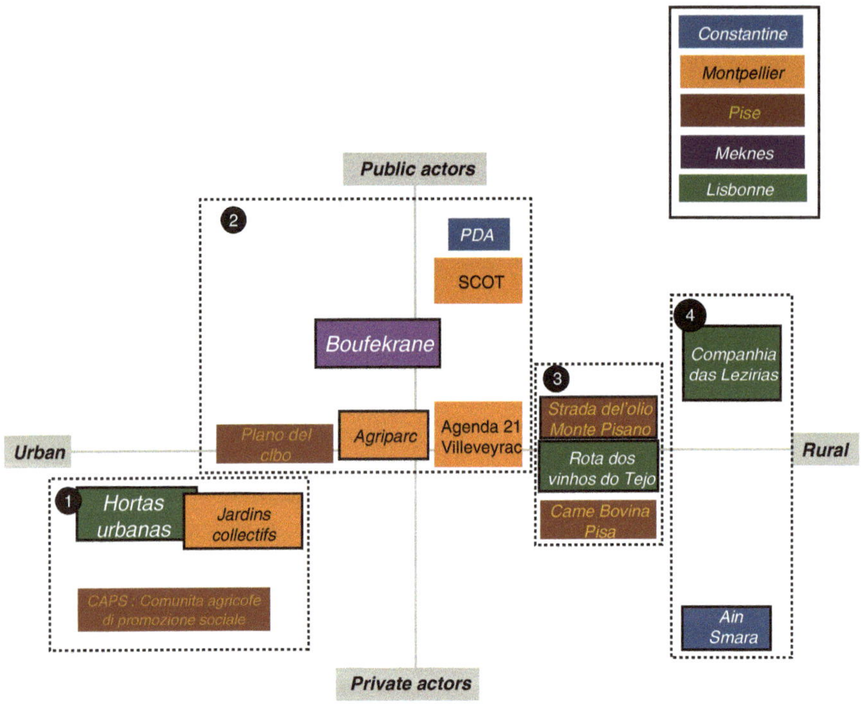

PDAU: Plan directeur d'amenagement et d'urbanisme - SCOT: Schema de coherence territoriale

Fig. 8.2 Illustration of the local projects identified on the five Mediterranean terrains

[2] The number of interviews conducted on each terrain varies according to the nature of the projects (e.g. the number of active actors, or the spatial scale). The method used consists of contacting a complete range of the protagonists in each case in order to gather the information necessary to characterize its course from emergence to the date of the investigation. Between 10 and 20 interviews were conducted in each case.

[3] For an example of the approach, see Vitry (2014) or Mousselin and Scheromm (2015).

The structure of this illustration facilitates the organization of the projects for the analysis of their internal dynamics. We have empirically grouped the projects shown in Fig. 8.1 among four families of projects: (1) projects initiated by private actors, often through associations, but more or less taken over by the urban public actor (e.g. allotment gardens); (2) projects conducted by public actors linked with other actor networks (either private or public); (3) projects emanating from local private actors, but supported by public actors; and (4) projects conducted by private actors, either organized or unorganized.

In the following, we analyse the processes of integration of cities and agriculture through eight projects. We present them in the same order as the groups identified in Fig. 8.1.

8.4 Integration Through Socio-spatial Practices: The Allotments Gardens

The development of allotment gardens was identified in Montpellier and Lisbon (see also, Chap. 12 by Scheromm and Mousselin).

Montpellier and Lisbon are characterized by the presence of collective municipal gardens within the city limits. In these cities, both civil society – citizens and associations – and the municipality play a strategic role in the creation and life of these gardens. Since 2004, pressure from local citizens has encouraged the municipality of Montpellier to develop a strategy for creating collective gardens: family gardens, parcels allocated to individuals for gardening, and shared gardens, cultivated collectively by neighbourhood residents. There are also natural spaces organized informally by urban gardeners. In Lisbon, we find two categories of collective gardens: unplanned gardens instigated by residents, generally in socially precarious neighbourhoods, and planned gardens that have been created in the framework of a municipal project aimed at developing urban agriculture in the city.

The dynamics involved in the creation of gardens are readily observable in Montpellier and Lisbon, but the issues surrounding these gardens are more diverse. In Lisbon, the large number of unplanned gardens ("squats" on municipal or private parcels) are symptomatic of the need of citizens to cultivate for food security. By contrast, this type of garden is not in evidence in Montpellier where the rapid development of shared gardens mainly reflects a desire for social connections and a setting, which satisfies the need to cultivate the soil as a means to maintain contact with nature. The various municipal actors cite these issues, but concerns over food security do not appear in the discourse of citizens or city officials. In the two cities, urban food production is certainly present, but for different reasons. The municipalities encourage production following the principles of organic agriculture, and thus contribute in some measure to the advance of agricultural practices that respect the environment in the city (Scheromm 2015), although sometimes with difficulty, particularly in Lisbon.

Urban collective gardens represent the spatial integration of agricultural spaces within the built fabric, as well as agricultural practices in cities. The urban actor (gardener, municipality) invents a space where cultivation is re-established within the "walls" of the city. The urban resident becomes a "farmer", in the sense that he practices a form of agriculture even without necessarily focusing on food production or commercialization (Soulard 2014). This is integration on a micro-local scale, through a practice that favours the proximity between agriculture and cities both spatially and socially.

This socio-spatial integration through small-scaled objectives, close and visible, provides a base for projects associating cities and agriculture, built and cultivated spaces. Some types of urban gardens enable a multi-generational social mix that may evolve naturally or be encouraged intentionally. They can enhance sensibility to environmental concerns and organic models of cultivation. They provide food and economic functions with objectives of food security and revenue. In addition, they are sometimes recognized politically in municipal planning documents and governance mechanisms (Mousselin and Scheromm 2015). In terms of the theoretical framework developed by Ruiz-Tagle, these socio-spatial processes of integration put into play the spatial and functional proximity dimensions of accessibility to resources. However, they are not exclusively intentional, nor are they invariably integrated into public action.

In conclusion, this is urban agriculture engaged by residents of the city, independent of professional, periurban and rural agriculture. The integration is vertical, among private and public actors, and observed as integration at the centre of urban activities and space dedicated to a specific form of agriculture where the citizen becomes an "agricultural practitioner" (Soulard 2014).

8.5 Integration Through a Pivotal Actor at the Interface: Private or Public Enterprises

Two initiatives emanating from agricultural enterprises were identified in Constantine and Lezíria de Tejo (near Lisbon).

Near Constantine, Algeria, the *Ain Smara* project is a farm, which has experienced both economic and agronomic transformations, and has influenced the local dynamics linking the city of Constantine with its neighbouring agriculture, traditionally dominated by cereal production. Based on diversification in lentil production, the project has introduced a collective agricultural dynamic. The growth and further development of the project throughout the region is supported by institutional actors (granting annual subsidies from the State to lentil growers), and partnerships with foreign companies encouraging the introduction of new cereal varieties and legumes for their production potential. Several farmers with allocations of public land in the region subscribe to the project to ensure continued access to their farms (Bendjaballah 2013) which benefit from the proximity of urban markets.

The *Companhia das Lezírias* is a Portuguese State-owned agro-forestry enterprise. It has land holdings estimated at almost 20,000 hectares adjacent to the Lisbon Metropolitan Area. The principal products of this region are grain for livestock, livestock production, rice, wine, and cork oak. Part of these lands is located in the National Reserve of the Tagus Estuary, an environmentally sensitive zone, important for its ecology and fauna. The principal objective of this enterprise is to insure the "profitable and efficient" management of this "agricultural and environmental public patrimony" (Accounting report, 2007–2012). The region includes a bird observatory, a farm for educational programs, an internationally renowned equestrian centre, and rural tourist lodging. A diverse collection of actors is involved in the *Companhia*. The enterprise is linked to different levels of Portuguese government and administration, local political entities, and other actors and professionals of the region. It is also affiliated with several organizations and associations, placing it at the intersection of different actor networks.

As we can see in these two examples, a single actor can act as an intermediary or "pivot" in the integration of a city and agriculture by tying together the different categories of actors: urban and agricultural, private and public. Such an actor possesses the resources required to exert different forms of leadership, and in Lisbon, the *Companhia das Lezírias* provides a useful example. This State enterprise "*is at the interface between metropolitan Lisbon and the rural world, and has a major pedagogic benefit for our visitors*".[4] Indeed, its location adjacent to the Metropolitan Area presents more than one face: that of a buffer zone against metropolitan urbanization which has suffered in the context of the on-going economic crisis (the abandonment of large construction projects); and that of a "*green lung*" offering the possibility to develop the advantageous proximity of nature and the city. In addition, the *Companhia* plays a significant economic role "*because it is at the interface between the city and the countryside which means that the markets can offer quality local products*".[5] Similarly, the Algerian *Aïn Smara agricultural* project, instigated by an individual actor is in a position to establish links with local farmers in order to develop local lentil networks, and encourage agricultural diversification in a region dominated by cereal production. This intermediary actor also brings together agricultural and institutional actors.

In conclusion, these city-agriculture relationships function through the private and public actors that are located in the metropolitan area. The new opportunities offered by cities (e.g. increased tourism in Lisbon or an emerging urban market in Constantine) reinforce agricultural or rural strategies. As opposed to the case of urban gardens, this represents a vertical integration in the sense that the urban logistics influence the functioning of agricultural strategies. Although the process of socio-spatial integration is not the particular objective of these projects, some actors can nevertheless take advantage of the compatibility of urban agricultural systems with the objectives of accessibility to collective resources. In this type of process, these are the coexisting functional dimensions, in the sense of Ruiz-Tagle, that are the most visible.

---

[4] Interview with *Companhia* actor.

[5] id.

8.6 Integration Through Land-Use Planning: A Non-build Valley and an Agricultural Park

Two initiatives emanating from urban policy were identified in Meknes and Montpellier.

In Meknes, the Boufekrane is one of three Wadi forming a natural valley that slices through the heart of urban space. This valley extends over a surface of 900 ha, and the intra-urban portion harbours diversified agricultural activities (market farming, arboriculture, and livestock). The Meknes Urban Agency's recognition of the potential role of this non-constructible space within the city has led to deliberations for the possible development of the valley, including an agricultural component. A project was initiated in 2009 with the objective of developing the "potential agricultural value for arboriculture and market gardening, and its unique landscaping qualities" which are essential to the "promotion of the region and tourism." (Agence Urbaine 2008). This represents an effort to draw attention to the benefits of the social functions, productive and environmental, of urban agriculture in a perspective for the valley that is open to the public. The project brings together different actors from the urban sphere (Urban Agency, government housing representatives, local officials, an association of real-estate promotion), but also agricultural actors from the DPA[6] and the Chamber of Agriculture. However, the project proposes an idealized vision for what urban agriculture could be, and none of the existing farmers in the valley has participated in the conception of the project. At this time, the project remains in the field of political intentions.

In Montpellier, an agricultural park in the upper valley of the Lez river is integrated into a mechanism of urban planning developed by the governing entity of the metropolitan scale (today, the Montpellier Méditerranée Métropole). This plan, known as the Schéma de Cohérence Territoriale (SCoT), was approved in 2006 and presents the agri-park as a strategic regional site to maintain and coordinate the viability of agriculture with the proper management of the river, and its riparian and leisure functions. This project long remained in the stage of zoning; however, in 2011 the Montpellier "Métropole" adopted an Agenda 21 to support sustainable development. The framework of this program organizes the objectives of the agri-park around four functions: agricultural production; local food supplies; the environment and landscape; and recreation and education. The agri-park has since grown in stature with the Métropole's acquisition of an estate, the domaine de Viviers (190 ha, including 110 cultivated hectares). Numerous actors have been involved in the project: the community, the landowners, SAFER (body mandated to find, organize and oversee contracts for land among farms), and the sixteen farmers renting the land. These actors have been joined by an agricultural cooperative, Terracoopa, whose mission includes the support of new organic farmers, engaged with short food supply. The project has provided the opportunity for certain farmers to enlarge their farms, while other new farmers face difficulties to develop their

[6]Agricultural services from the government agency, *Direction Provinciale de l'Agriculture* (DPA).

activities, and urban actors struggle to improve management methods in the agri-park.

Both of the cases studies, the non-build valley and the agri-park, show models of landscape management based on the identification of urban and agricultural issues in the design and operations of a multifunctional space. The integration takes place in the formation of actor relationships and in the access to community resources, the combination of which theoretically allows the construction of a hybrid space and identity, both urban and agricultural. Nevertheless, the relationships between actors are not necessarily symmetric. The cases developed include public action instruments that are difficult to translate into governance mechanisms capable of interacting with various heterogeneous components.

In conclusion, the relationships linking city and agriculture in these projects begin with the investment of urban actors committed to agricultural spaces, and operate through the process of interaction between institutional and professional actors. The disparity of expectations carried by the various actors and the difficulties inherent in cooperative efforts still dominate the emergence of these relationships, but new links between cities and agriculture are being formed and tested.

8.7 Integration Through Flows and Mobilities: Wine and Olive Oil Routes

Two initiatives emanating from rural policy were identified in Portugal and Pisa.

In Portugal, the wine route of the Tagus river valley (Lisbon) was created in 1998 as a means of expanding cultural opportunities, gastronomic and touristic, with wine as the focal point. The resulting tour is composed of four circuits across a region of more than 6000 km^2. Twenty-four local and regional actors participated in the creation of the project (farmers, cooperative wine makers, regional entities responsible for tourism, a regional wine commission of Tagus valley wines). They were organized in an association called Associação da Rota dos Vinhos do Tejo which was created to coordinate and manage the route. This project represents a regional network organized around the wine industry. Tours and meeting places are indicated with road signs installed in 2002, but they are also promoted through various activities and events, and on internet sites.

In the urban region of Pisa, Italy, an olive oil route was formally established in 2005 with financial support through the EU's rural development policy. For a few years prior to this, the route known as La Strada dell'olio Monti Pisani existed in the form of an informal consortium, until the Tuscan region legislated the recognition of the route as a mechanism to promote the production of olive oil. With a format similar to the wine route of the Tagus valley, La Strada dell'olio is a touristic tour developed to explore this region known for its olive oil production, but the primary objective was the support and promotion of agricultural production, and efforts to avoid the abandonment of olive groves in an area north of Pisa. The olive oil route currently unites 28 associates: municipalities, farmers, olive oil producers, representatives from the agri-tourism and hotel industry, cultural and nature associations, and food outlets. Most of these actors are located in the periurban and rural hills known as Monte Pisano, but they are also found in the Pisa metropolitan area. The different actors are organized in an oversight committee that is responsible for the coordination, management, distribution and promotion of regional products (e.g. marketing, annual festivals, research and advertising).

Even if regional networks for marketing and distributing food and agricultural products were created as a means of economic support and rural development, they can also play a role in the integration of a city and its agriculture. The thematic "routes" observed contribute to the flow in the form of the circulation of actors and their resources. They reconstruct the links between production and consumption, and the supply networks regionalize and anchor the links between the city and its nearby agriculture, previously more oriented towards exterior markets. In Lisbon, the wine route becomes a privileged mechanism to advance the integration of the city and regional agriculture by enabling the relationships between agricultural and urban actors through the market for tourism. In Pisa, the olive oil route also works as a mechanism that goes beyond the promotion of products, supporting the region through tourism and enabling the emergence of a collective action (initially within the perimeter of Monte Pisano): the creation of a collective of farmers, and connections with the public sphere. The processes of integration between the city and agriculture are an unintentional result and can be observed on several scales. The preservation of olive groves protects the city of Pisa from environmental risks (e.g. weather-related flooding). The route increases the visibility of an agriculture product reinvested gradually by the urban gastronomic culture. The collective action organized around agriculture encourages participating municipalities to develop initiatives that create the common norms and standards for dialogue with urban actors in the framework of development mechanisms associating the city and its neighbouring rural agriculture.

In the conclusion, the creation of agricultural routes provides an opportunity for farmers that have little previous experience organizing on such a scale, and limited connections with urban actors. It is also a tool that has facilitated the links between municipal actors that have rarely collaborated, particularly at the level of development projects in their respective regions, or in interaction with neighbouring cities.

8.8 Conclusion: Analysis of the Integration of Cities and Agriculture

Several processes of integration can be observed in the eight local projects previously presented. It is clear that the socio-spatial integration of cities and agriculture is not an intentional objective in the majority of the local projects we have observed. An on-going process more or less operates through the practices, the actors and the exchanges that take place among them. The projects that do incorporate this objective at their conception are predominantly institutional in nature; the difficulties noted in their development reflect the challenge of managing these complex processes, and the influence of competing power strategies. The integration of cities and agriculture appears to be more a matter of transition over the long term, following a complex and hesitant path.

Following the theoretical proposition of Ruiz-Table, the diverse forms and processes associated with integration projects show that socio-spatial integration manifests itself as a multidimensional relationship that functions independently and on different scales. These processes can have both "positive" and "negative" effects.

At the micro-scale of integration through practices, the successful projects show that the most significant dimensions are those of geographic proximity and accessibility to resources. Thus, we can expect this integration to result in the reduction of spatial and social divide between cities and agriculture. Nevertheless, more negative effects can emerge at this scale as well, such as the barriers of exclusion or situational conflicts that have been identified in the literature on urban gardens (Barthel et al. 2010). At the meso-scale, we can identify two additional categories of effects resulting from the integration processes that operate through the flows and networks. First, there is a socio-economic effect when one category of actors takes advantage of urban proximity to create an economic resource. By integrating the urban components into agricultural strategies, these actors capture the value of the marketing potential provided by this urban proximity. Secondly, experience through the collaboration between private and public, urban and rural categories of actors, often reveals various capacities for innovation and the construction of an enduring collective spirit. On this scale, coexistence and the construction of a shared vision are the most obvious dimensions of the socio-spatial integration processes.

In conclusion, our analysis shows the diverse spatiality and multi-dimensionality in the processes integrating cities and agriculture in these Mediterranean regions. However, the depth of complexity in these processes is far from exhausted. There remains a need to further explore the degree of integration realized through local projects. This suggests analysing projects, and the processes involved, across an expanded grid of more diversified Mediterranean regions, and over more extended periods.

Acknowledgement This research has been funded by the French National Agency for Research. DAUME project under n° ANR-2010-STRA-007-01.

References

Agence Urbaine de Meknès (2008) Elaboration du plan paysager de la vallée de l'oued Boufekrane. Etude de mise en valeur, Mission II, Rapport définitif, 48 p

Barthel S, Folke C, Colding J (2010) Social–ecological memory in gardening: retaining the capacity for management of ecosystem services. Glob Environ Chang 20:255–265

Bendjaballah OB (2013) Politiques urbaines, terres agricoles et marché foncier: quel avenir pour l'agriculture périurbaine à Constantine (Algérie) ? Cah Agric 22(6):544–551

De La Salle JM, Holland M (2010) Agricultural urbanism: handbook for building sustainable food & agric systems in 21st century cities green frigate books. Distributed by Independent Publishers Group, 200 p
Ernwein M (2014) Framing urban gardening and agriculture: on space, scale and the public. Geoforum 56:77–86
Giacché G (2014) L'expérience des parcs agricoles en Italie et en Espagne: vers un outil de projet et de gouvernance de l'agriculture en zone périurbaine. Géocarrefour 89(1–2):21–30
Gilly J-P, Torre A (dir) (2000) Dynamiques de proximité. L'harmattan, Paris, 302 p
Jarrige F, Thinon P, Nougaredes B (2006) La prise en compte de l'agriculture dans les nouveaux projets de territoires urbains. Exemple d'une recherche en partenariat avec la Communauté d'Agglomération de Montpellier. Revue d'Économie Régionale Urbaine 3:393–414
Kühn M (2003) Greenbelt and green heart: separating and integrating landscapes in European city regions. Landsc Urban Plan 64(1–2):19–27
Lee R (1993) Integration. In: Gregory D, Johnston R, Pratt G, Watts M, Whatmore S (eds) The dictionary of human geography, 3rd edn. Wiley-Blackwell, Malden
Loudiyi S, Maury C, Lardon S (2011) Agriculture(s) et projet urbain durable: issues incertaines, destinées imprévisibles. VertigO – la revue électronique en sciences de l'environnement 11(2). https://doi.org/10.4000/vertigo.11481
Marraccini E, Lardon S, Loudiyi S, Giacché G, Bonari E (2013) Durabilité de l'agriculture dans les territoires périurbains méditerranéens: enjeux et projets agri-urbains dans la région de Pise (Toscane, Italie). Cah Agric 22(6):517–525
Mousselin G, Scheromm P (2015) Vers une mise en politique des jardins collectifs urbains: Approche comparée de deux trajectoires municipales à Montpellier et à Lisbonne. Articulo Special issue 6. https://doi.org/10.4000/articulo.2685
Niwa N (2009) La nature en ville peut-elle être agricole? De la Suisse au Japon. Urbia: les cahiers du développement urbain durable 8:103–126
Paoli JC, Soulard C (2003) Note de méthode du projet inter-unités RIDT « réseaux, interactions et dispositifs territoriaux ». Comment écrire la chronique d'un dispositif territorial ? Séminaire RIDT, INRA-SAD, Dijon, 12 p
Paül V, McKenzie FH (2013) Peri-urban farmland conservation and development of alternative food networks: insights from a case-study area in metropolitan Barcelona (Catalonia, Spain). Land Use Policy 30(1):94–105
Perrin C, Jarrige F, Soulard C-T (2013) L'espace et le temps des liens ville-agriculture: une présentation systémique du cas de Montpellier et sa région. Cah Agric 22(6):552–558
Poulot M (2011) Des arrangements autour de l'agriculture en périurbain: du lotissement agricole au projet de territoire. VertigO – la revue électronique en sciences de l'environnement [Online] 11(2). https://doi.org/10.4000/vertigo.11188
Poulot M (2014) L'invention de l'agri-urbain en Île-de-France. Quand la ville se repense aussi autour de l'agriculture. Géocarrefour 89(1–2):11–19
Ruiz-Tagle J (2013) A theory of socio-spatial integration: problems, policies and concepts from a US perspective. Int J Urban Reg Res 37(2):388–408
Scheromm P (2015) Motivations and practices of gardeners in urban collective gardens the case of Montpellier. Urban For Urban Green 14(3):735–742
Scheromm P, Perrin C, Soulard C-T (2014) Cultiver en ville... Cultiver la ville ? L'agriculture urbaine à Montpellier. Espaces et Sociétés 158(3):43–66
Soulard C-T (2014) Pratiques, politiques publiques et territoires: construire une géographie agricole des villes. HDR report, Université Bordeaux-Montaigne, 188 p
Specht K, Siebert R et al (2014) Urban agriculture of the future: an overview of sustainability aspects of food production in and on buildings. Agric Hum Values 31(1):33–51
Tornaghi C (2014) Critical geography of urban agriculture. Prog Hum Geogr 38(4):551–567
Torre A, Zuindeau B (2009) Les apports de l'économie de la proximité aux approches environnementales: Inventaire et perspectives. Nat Sci Soc 17(4):349–360

Valette E, Philifert P (2014) L'agriculture urbaine: un impensé des politiques publiques marocaines? Géocarrefour 89(1–2):75–83
Viljoen A (ed) (2005) CPUL continuous productive urban landscape. Architectural Press, Oxford
Viljoen A, Bohn K (2014) Second nature urban agriculture: designing productive cities. Routledge, New York
Viljoen A, Wiskerke JS (eds) (2012) Sustainable food planning: evolving theory and practice. Wageningen Academic Pub, Wageningen
Vitry C (2014) Vers une théorie des apprentissages de gouvernance: une exploration en territoires périurbains. PhD thesis, Montpellier SupAgro (Unpublished)

Part III
Inventory of Changes in Urban Agriculture

Chapter 9
Innovative Commercial Urban Agriculture in the Paris Metropolitan Area

Christine Aubry and Anne-Cécile Daniel

Abstract For centuries, Paris and its Region were hot-spots of innovative agriculture, chiefly in market-gardening production. In the past 5 years, as part of the global movement reconnecting consumers with food production, innovative commercial forms of urban agriculture have been emerging inside the city Through systematic surveys, regular meetings and direct involvement in the experimental design of new cropping systems on rooftops, we have studied the growing number of UA projects in Paris and its region. Three types of project can be distinguished regarding their relationship to cropping support: on-the-ground farms, "low-tech off-soil" systems with open-air cropping systems on rooftops, and "high-tech" greenhouse systems. Project leaders are from widely diverse backgrounds, generally have no agronomic skills, have a high level of academic education, and are very much in tune with the city way of life. Our first estimations may be used to measure the possible contribution of these forms of Urban Agriculture to the city's food supply – a contribution that is still quantitatively small. Parisian and regional authorities are now supporting this emergence in various ways. Yet these projects still face some challenges, including difficulties in securing access to open spaces or buildings, and the current suspicion about the safety of agricultural products grown inside reputedly polluted urban region.

Keywords Urban agriculture · Commercial projects · Project leaders · Agronomy

---

C. Aubry (✉) · A.-C. Daniel
INRA, AgroParisTech, University of Paris-Saclay, UMR SADAPT, Paris, France
e-mail: christine.aubry@agroparistech.fr; anne-cecile.daniel@agroparistech.fr

C.-T. Soulard et al. (eds.), *Toward Sustainable Relations Between Agriculture and the City*, Urban Agriculture, https://doi.org/10.1007/978-3-319-71037-2_9

9.1 Urban Agriculture in Paris Region: An History of Innovations and Fluctuations

Innovations in urban agriculture in and around Paris are … not so new! The first tests to improve horticultural techniques (varietal selection, pruning fruit-trees, invention of cold frame, etc.) were developed in the seventeenth century in the "Potager du Roy" near the Versailles Palace, by the Royal Gardener, La Quintinie (Quellier 2003). These experiments were continued and furthered during the nineteenth century by local market gardeners to meet the increasing demand for vegetables from the growing Parisian population which valued quality, freshness and diversity. The size of most of the market gardening farms was between 5000 and 10,000 m^2, with five or six full-time workers (Moreau and Daverne 1845). Their motto was "cultivate the smallest garden, but as well as possible". New techniques allowing gardeners to supply the market throughout the year were tested and used extensively, such as "hot layers" with horse manure, lettuce under glass bells to obtain early harvests, and so on. At the end of the nineteenth century, the Paris metropolitan Region was one of the most innovative and productive market gardening areas in the world (Roy 2012), and these techniques still inspire innovative gardeners worldwide (Coleman and Petit 2013). The number of market-gardeners and their localisation show that in Paris and its surrounding areas significant changes occurred: from the mid-nineteenth century to 1912, their number increased following the growing demand for fresh products, but chiefly outside Paris (Phlipponeau 1956); up to the Second World War, the global number of market gardeners decreased, but it was after 1945 that market gardeners leave inner Paris intra-muros, dramatically decreased in its vicinity while the surroundings areas and remote supply areas played a growing role in supplying the city (Roy 2012; Brunet and Charvet 1994) The globalization of food supply chains during the last quarter of the twentieth century (Morgan et al. 2006) led to another fast decrease in the number of local market gardeners.

Throughout history, the relationships between Paris and its neighbouring areas have fluctuated in terms of food supply. Although Paris has never been self-sufficient (Abad 2002), the participation of peri-urban and intra-urban agriculture in its urban food systems, and the nature and strength of relationships, between the city and its agriculture of the city and that of its surrounding areas, and specifically between local farmers and urban consumers have always varied widely (Poulot 2015). During and just after the Second World War, many urban spaces were used for agricultural purposes, but the second half of the twentieth century gradually witnessed a relative disconnection between the region's food supply and the majority of the regional farming systems, largely oriented to the worldwide market for arable crops or to vegetable exports (Poulot and Rouyres 2000). One of the focal points was the destruction in 1971 of "Les Halles" (Paris' wholesale market that had been in the centre of the city since the Middle Ages), and the construction, outside the city, of the Rungis wholesale market. As a central node in an international network, Rungis supplies a large part of the Parisian region's fruit, vegetables, meat etc. It rapidly became the leading European platform for fresh produce sales. In 2004 a specific

Regional Producers' Floor was created inside the Rungis' Wholesale market, but only for professionals such as restaurants.

The late twentieth and early twenty-first centuries have however witnessed a reactivation of links between local agriculture and consumers, mainly though the wide-scale development and diversification of short supply chains in the Ile-de-France Region (Aubry and Kebir 2013). This phenomenon follows a more national movement of reconnection between urban consumers and farmers (Aubry and Chiffoleau 2009), largely due to urban consumers' growing defiance of the "globalized agri-food system" (Morgan et al. 2006). This is partly a result of the various sanitary crises (mad cow disease, avian flu, etc.) during the 1990s. Many organisational innovations have taken place in recent years to re-invent relationships between producers and consumers (Redlingshöfer et al. 2014). But even though agricultural areas still account for nearly 50% of the region's surface area, the number of market-gardeners is consistently decreasing. Two-thirds of all market gardening farms ceased to exist between 2000 and 2010 (Agreste 2013).

A new phenomenon has nevertheless been emerging in the past 5 years in Paris: innovative forms of agriculture within the city, which are at least partially business-oriented. Some are inspired by experiences that already exist in North America (Cohen et al. 2012) and/or in Asia (the firma Sky Greens for instance) and a majority of them are not yet realized. What form do or will they take in Paris? What are their objectives and the profiles of the project leaders? How do (or could) they contribute to the urban food systems? What are the drivers of and the obstacles to this "re-introduction" (from an historical point of view) of agricultural forms inside the city of Paris?

To help answer these questions, an on-going research project has been designed. The entirely non-professional forms, such as collective gardening, are not considered here, despite their very rapid growth (from 5 in 2003 to 124 in 2013 in Paris itself, Pourias, 2014) and the fact that some of them, at regional level, may make a relatively significant contribution to the vegetable supply of gardeners (Pourias et al. 2015a). We will focus here on emerging commercial forms of urban agriculture (below UA), which are growing too, but are just beginning to be researched. These forms were approached by combining diverse methods:

(i) An annual meeting with the intra-urban project leaders: since 2012, we have been organizing an informal meeting between project leaders and our team once a year, where the project leaders exchange their experiences, aims and obstacles. The invited project leaders re were six in 2012 (5 attending), 17 in 2013 (14 attending) and 32 in 2014 (19 attending). Our invitation list included all the at least partially commercial projects with which our team was in contact. These regular meetings allowed us to be able to progressively build an initial typology of these projects (see below).
(ii) A questionnaire with ten questions was sent to the 32 invited project leaders in 2014.We received 19 answers and we were able to complete three additional surveys thanks to personal contacts to reach 22 answers. This on-line questionnaire includes two parts, one about the personal data of the project

leader (age, professional skills, studies level, native region), and one about the project itself (company status, main activities, progress of the project).

Another important point for this relationship with Project leaders is that our team is itself a recognized conception point of urban agriculture systems in Paris. Since 2012, we have been involved in the design and evaluation of roof-top gardening cropping systems (Grard et al. 2013, 2015) which has led us to develop deeper contacts with these Project leaders, to share preoccupations with them and even to contribute to their knowledge regarding choices of crops, crops successions, technical management and pollution problems mainly in the context of rooftop gardening.

In this chapter we present: (i) the different forms of these innovative commercial urban agriculture projects in Paris and the profiles and skills of the project leaders (ii) the available knowledge and questions about their possible contribution to the urban food system; followed by (iii) a panorama of the drivers of these projects as well as some of the main obstacles to their development.

9.2 Diversity of Commercial Urban Agriculture Forms and Their Project Leaders

Various typologies of UA in Paris can be defined (Fig. 9.1). As Paris is one of the densest cities in Europe (21,300 hab.km^{-2}), with scarce "on the ground" spaces for cultivation, the technical criteria of the type of technical systems used, regarding access to the city's "ground", seem to be relevant.

9.2.1 A First Typology of Commercial UA Projects

On the basis of our own contacts and bibliography, a first typology of the commercial UA projects has been proposed (Daniel 2013; Daniel et al. 2013) based mainly on technical choices (Table 9.1). We thus distinguish: (i) small urban farm projects on urban soil, in situations where urban soil is not too scarce or too polluted; (ii) "low-tech" farm systems, off-ground, which use technosols or soils based on waste (with or without transported soil from rural areas); and (iii) "high tech" urban farms projects, mainly based on hydroponic or even aeroponic and aquaponic production systems in greenhouses, which could include an integrated vertical farm project in the future (The "Romainville Market Vegetable Tower" foreseen for 2018). It already includes an innovative production of mushrooms on coffee grounds (Fig. 9.2). The latter two cases of off-ground systems may be considered as forms of "ZFarming", for "Zero acreage farming", as defined by Specht and Thomaier (Specht et al. 2013; Thomaier et al. 2014).

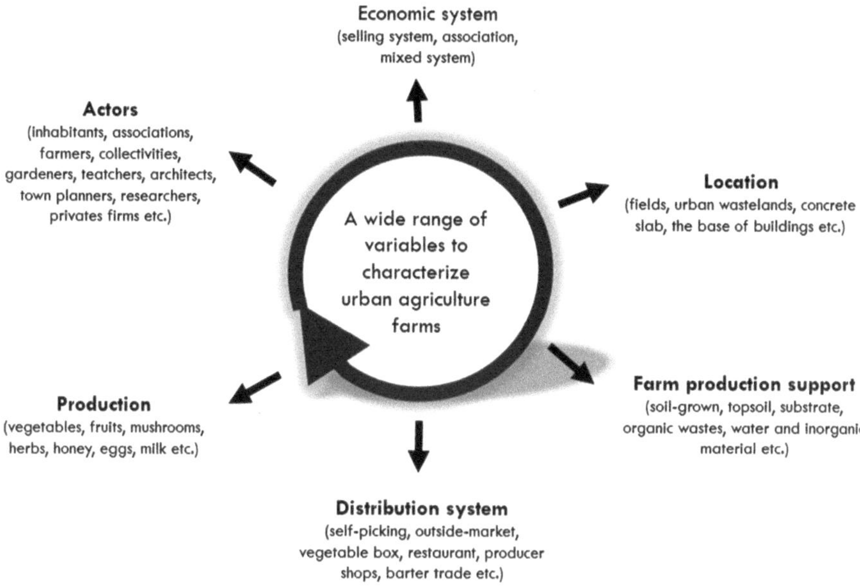

Fig. 9.1 A wide range of variables to characterize urban agriculture farms (Source: Daniel 2013)

Table 9.1 Number of projects according to the "technical system" typology

1-Urban "on-the ground" projects	2-Low tech off-ground projects	3-High tech urban farms
6	8	8
1-A: 1000 m²; SU	2-A: 300 m²; IP	3-A: 150 m²; SU
1-B: 2 ha; SU	2-B: 100 m²; IP	3-B: No realization; vertical farming; SU
1-C: 1300 m²; SU	2-C: 3 ha; SU	3-C: No realization; GH on rooftop; IP
1 D: 600 m²; IP	2-D: 100 m²; IP	3-D: No realization; vertical farming; SU
1-E: 650 m²; SU	2-E: 150 m²; IP	3-E: 150 m²; IP
1-F: 450 m²; IPs	2-F: 2000 m²; SU	3-F: No realization; AQ; SU
	2-G: 40 m²;IP	3-G: No realization; AQ; SU
	2-H: 1000 m²; IP	3-H: No realization; GH on rooftop; IP

IP inner Paris, *Su* suburb, *GH* greenhouse, *AQ* aquaponic

The first type is "on-the-ground" inside the city. This form is current in North American cities such as New York and Detroit, due to an urban structure that is less dense than in Paris and, in the case of Detroit, large-scale de-industrialization that has led to numerous vacant lots re-used for agricultural activities (collective gardens, commercial urban farms (Cohen et al. 2012; Mc Clintock et al. 2013; Paddeu

Fig. 9.2 An innovative farm producing mushrooms on coffee grounds: the U-Farm (with the kind authorisation of C Péchard UFarm)

2012, 2015). In the Parisian region, there are traditional farms in peri-urban areas, as well as some innovative forms of urban agriculture on the ground. Five projects of "on-the-ground urban farms" are located in the very close suburbs of the city (which will be included in next years planned metropolitan development of 'Le Grand Paris'), but in 2014 a new project emerged in inner-city Paris in the Bois de Vincennes green park. These urban farms are small (a few thousand square meters to less than two hectares) but at least some land is dedicated to crops that will be sold through "local boxes schemes", or direct on-farm selling, or else to a solidarity restaurant nearby. The economic models of these farms are of course mixed: apart from the income from sales, the farms are also recognized as multifunctional urban projects (Zasada 2011; Poulot 2013), with a part of the area dedicated to collective gardening and educational or cultural events for the city. A part of their income is therefore from remuneration of these services. They are also supported by organizations that advocate for sustainable development and transition cities, and the cities councils are willing to support them for neighbourhood rehabilitation and reactivating social relations. Frequently subsidies from the city council contribute more to the funding than does the farm's business activity. It is the case for five out of our six cases belonging to this first model. Conscious that these subsidies will certainly decrease with the economic crisis in the net future, all these "on-the-ground" urban micro farms aim to develop their own incomes, through an increase of products selling and better remuneration of educational/leisure services.

The second type of commercial agriculture is "off-ground", mainly open-air and adapted to rooftops, for instance. Although some collective rooftop gardens in Paris and the suburbs use soil transported from arable farms in the region or from building sites, other innovative projects have created a new type of soil, on the model of the experimental rooftop garden of the School AgroParisTech. This experiment was built to design cropping systems using only urban organic substrates (such as green composts, crushed wood and coffee grounds with mushroom mycelium, cropping residues of a container urban farm). The experimental data on production and on pollution risks, obtained on the AgroParisTech rooftop over a two-year period (Grard et al. 2013, 2015), are now one of the main bases for designing these types of open air rooftop gardening in Paris. A young company, Topager©, whose leaders participated in the launch of the AgroParisTech rooftop experiment, now develops such rooftop gardening for customers, both in the private sector (gastronomic restaurants, firms) and public institutions or non-profit organisations (community centres, therapeutic gardens for disabled persons). Generally, customers either want to rent their rooftop or their terrace to an urban farmer, or else prefer to lend a space for social uses such as collective gardens, as a form of "green image". In the latter case, they employ a facilitator or have a maintenance contract with specialists such as Topager©. Eight "open-air off-ground farming" projects existed at the end of 2014 in Paris and its suburbs, but none of them directly sold vegetables boxes. In the case of Topager© the business model excludes the sale of agricultural products; the company sells only the design, implementation and maintenance of rooftop gardens for its customers. The other projects, even if they have a commercial part, keep a certain level of multi-functionality, which is reflected in their various sources of income (rented collective gardens, sale of gardening supplies, cultural animation) as was the case for the first group.

The third type is entirely "off-ground" and mainly concerns rooftop greenhouse projects. To date, eight companies have been created. The emerging projects are mostly hydroponic, although two have aquaponic systems and one an aeroponic system. An original form of urban farm produces mushrooms on coffee grounds in a shipping container (see upper Fig. 9.2). It began in 2012 with a demonstration in Paris, but now the firm has developed a production space in the cellars of the Rungis' Platform and sells the mushrooms directly to gourmet restaurants. In these cases, the business model is entirely based on the sale of the products, and project leaders aim for high levels of economic profitability with this single source of income.

Behind this diversity of technical components and business models, all these types of innovative agriculture have three common points:

- They all try to operate within short supply chains, through specific links with gastronomy for example. All these projects promote the real or assumed benefits of ultra-short supply chains without any (or almost no) need for transportation of the vegetables: crops harvested at full ripeness, fragile species to rediscover, decrease in greenhouse gas emissions by avoiding transportation;
- they all claim to take into account social aspects, either directly by a part of their activities (chiefly the first type) and/or by participating in local social insertion

by employing local unemployed and/or disabled persons (second and third types);
- They all claim to participate in the urban metabolism (Barles 2009), for example by using urban waste as cropping substrates and by reducing the energetic costs of the production (reduction of the energy consumption of greenhouses by recuperation of urban heat, etc.).

9.2.2 *Profiles and Skills of the Projects Leaders*

The UA trend is growing today and various actors are taking a keen interest in this form of agriculture. We could directly testify to fact with the huge increase in project leaders coming to our annual meetings. We also very frequently meet emerging companies, architects, town planners, and students who want to develop projects or are interested in gaining a deeper knowledge of the phenomenon. Some of them have no clear ideas of what they could do, others have innovative ideas but no means to implement them, and a handful are setting up, or about to set up, a project. Some of them are not really UA "makers" but corporate sponsors or partners of start-ups.

Concerning only the 22 Project leaders who answered our questionnaire, we observe a variety of professional origins. The majority of innovative intra-urban UA entrepreneurs have no agricultural or agronomical background, only 4 on 22, which means less than a quarter, have studied agriculture or agronomy (Fig. 9.3). There is thus a large panel of skills, and specifically more and more people who have a commercial/marketing background. We also find city planners or architects, often in connection with sustainable development, and even social workers.

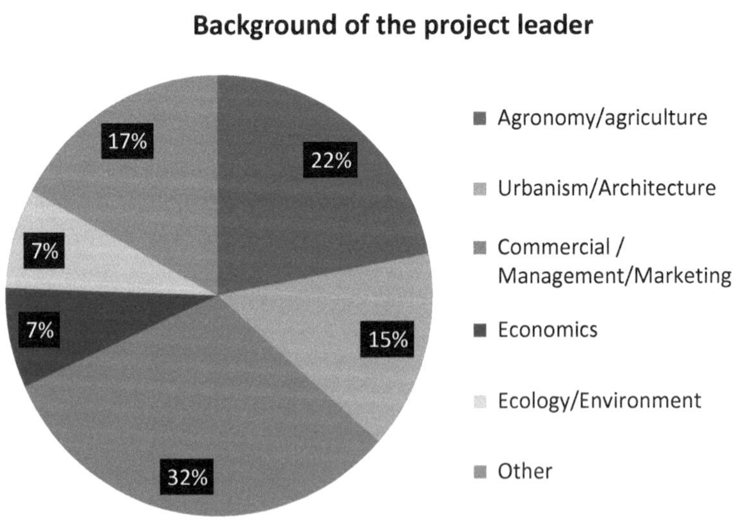

Fig. 9.3 The professional background of 22 intra-urban UA project leaders

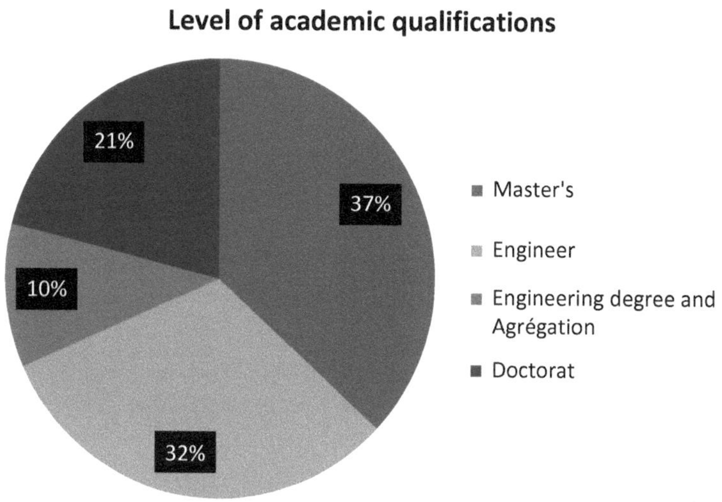

Fig. 9.4 Level of academic qualifications of 40 intra-urban UA project leaders

These Project leaders then have two things in common: an urban way of life and a professional background not linked to agriculture. A third significant shared factor is that they all have a high academic level (Master's at least) (Fig. 9.4). These stakeholders conceptualize these new forms of urban agriculture as they understand the various flows of water, waste, energy, biomass and so on generated by the city. They believe that UA is a tool to change the city lifestyle and is a source of innovation.

While many of these entrepreneurs are currently seeking to develop a business plan, they also need to integrate technical skills in the project, mainly in agriculture or horticulture. Their economic models are therefore very fragile. The projects encounter regulatory bottlenecks and lack technical expertise. This is specifically the case of the third category of project leaders, those who want to develop rooftop greenhouses and/or other "high tech" systems. The leaders of these projects are entrepreneurs who have skills either in construction, or in business, or in the energy industry. They have recruited or plan to recruit technical staff for vegetable production as they lack skills for horticultural cropping systems management, chiefly in highly specialized systems. Of the four agriculture/agronomy specialists involved in an UA project, whatever its form, two have created their own company or non-profit association, and the other two have been very recently (less than 2 years ago) recruited by investors or entrepreneurs. This proportion of technicians is likely to increase in 2015 and 2016 thanks to new ideas and interest of this new sector.

We posit that the strong inclusion of the project leaders in city life could facilitate today and in the future their capacity to detect and take into account in their commercial intra-urban systems the desires of urban inhabitants for a more local consumption (Aubry and Kebir 2013). They are thus aware that city dwellers would like to be more in contact with food and quality products through short (here "ultra" short) supply chains, to have the possibility for them and their children to learn about cultivation techniques, and to participate in a renewal of social links (Levidow et al. 2010; Pourias et al. 2015b).

For some project leaders who lack of technical skills, a possible problem in the future is that their lack of knowledge of the more "traditional" agriculture in peri-urban areas could lead to difficulties with the agricultural professional sector, whereas possible links and complementarities with other forms of agriculture may be necessary (see below).

9.3 What Contributions of These Forms of Urban Agriculture to Urban Food Systems?

The Parisian food supply is a complex system with little resilience, as it depends largely on the external supply, essentially by road transportation. It has been calculated that, should these supply lines be cut, the French capital would have only 2–3 days of food reserves (Toulallan 2012). This fact is a source of stress for the regional authorities. A recent study funded by the Regional Council (Toulallan op.cit) indicates that the total vegetable consumption (excluding potatoes) in the whole Ile-de-France Region is approximately 480,000 tons a year, for the 12 million inhabitants. Of this, 125,500 tons (26%) are from peri-urban horticulture. Inside Paris (2.2 million inhabitants), the global annual vegetable consumption is reportedly close to 90,000 tons. Thus, one of the first questions is how far these innovative forms of urban agriculture could contribute quantitatively to the Parisian food system. Of course, the diversity of these innovative projects leads to various potential elementary productivities, with variable possibilities of being duplicated.

There is a crucial shortage of quantitative data on these productivities. Only estimations are available, from on-site experiments and more frequently from documentary data. Regarding open-air rooftop gardening, the experimental project of AgroParisTech yielded (in lettuce and tomatoes for the first two experimental years) between 5 and 8 $kg.m^{-2}.year$ – for the most successful experiments (Grard et al. 2015). That means around 50 to 80 tons per hectare, which is over the local professional market gardeners' production levels in open-air systems. A first study by the Paris Municipality estimated that around 80 ha of flat rooftops could be "potentially cultivable" inside Paris (APUR 2013). This is an upper limit of cultivating capacity, as this study was based only on aerial photos and basic data on the buildings. Nevertheless, if the total 80 ha of "cultivable flat rooftops" in Paris were under cultivation, with the upper level of open-air rooftop production (around 80 $t.ha^{-1}$), it would lead to the production of around 6400 tons of edible vegetables a year, which is less than 10% of intra-urban consumers' needs. Moreover, we know that this estimate of Parisian flat rooftop cultivable surface area is over-optimistic, as it fails to take into account the real load-carrying capacity of the roof, its waterproofness, and the legal status of the building, all of which can constitute real obstacles. Moreover, according to Topager©, first experiences on different rooftop configurations showed that only 50 to 70% of a roof can effectively be cultivated, because of the area required for paths, technical equipment, etc. Then the potential contribution of Parisian rooftops under open-air technical systems to the food supply of the city would be small from a quantitative point of view.

Under hydroponic systems, the greenhouses would be more productive, even if the literature shows a high variability of annual production: around 20 kg.m^{-2} (Sanye et al. 2012) as opposed to over 50 kg.m^{-2} for such systems used in rural vegetable production. With the same potential rooftop surface area as in the previous cases, we would obtain 40,000 tons a year, which is significantly higher but not enough to ensure Paris' self-sufficient. Moreover, the technical, financial and socio-legal conditions for setting up a greenhouse on a rooftop are complex, making it probably more difficult than setting up a "simple" open-air rooftop garden. It is thus highly probable that only a very small part of the flat rooftops in Paris could be under cultivation through greenhouses systems.

Thus, from a quantitative point of view, these systems could contribute to Paris' vegetable consumption, but only to complete and not to replace the current food supply. Alone they will never "feed Paris". This result is consistent with other studies in the industrialized world, showing that the capacity of intra and peri-urban agriculture to contribute to cities' food supply may be significant but never sufficient (Porter et al. 2013; Morgan 2009). Nevertheless, as noted above, all these innovative forms of UA are promoting the development of short supply chains with local consumers, including on a very local scale, the scale of the concerned building and its immediate neighbourhood. This means that it could play a significant role for specific populations, from not only a quantitative but also a qualitative point of view, through the possibility of supplying very fresh products without transportation, allowing for the cultivation of old species and varieties, etc. The nutritional, educational and gustative aspects of these types of food supply could then be taken into consideration. As shown for collective gardens, the "food supply function" is far from being only a quantitative one (Pourias 2014). On these points as well, it will be important in the near future to take into consideration the fact that these innovative forms of urban agriculture are positioned more or less in the same market niche as the regional short supply chains: urban consumers interested by having a more direct food supply. Would the Parisian consumers be more interested by supporting new intra-urban producers than peri-urban farmers (who could even disappear if not supported)? Yet not all the vegetables consumed by Parisians are easy to grow in these innovative forms of urban agriculture, especially in greenhouses (for example tubers such as potatoes, carrots etc.). How and to what extent these new forms of UA may be in competition with and/or complementary to the short supply chains from peri-urban agriculture in the near future is an important topic which has to be explored. We already know that in Montréal (Canada) the well-known hydroponic greenhouse Lufa Farm has organized direct complementarities with peri-urban market gardeners, thus enabling it to complete the range for its customers, but the ways and modalities of these arrangements are for the moment unpublished.

9.4 Opportunities and Obstacles for Innovations in Urban Agriculture

The Parisian and regional authorities are increasingly interested in supporting these innovative forms of urban agriculture. Three recent types of initiatives have been taken.

In 2013, Paris Region Lab, a research team belonging to the Paris municipality, and the Urban Ecology Agency of Paris (AEU) launched a Call for Projects entitled Végétalisations Innovantes (innovative greening). The main purpose of this Call for Projects was to initiate and/or generate projects relating to UA and/or biodiversity in Paris and region. Végétalisations innovantes was engaged in supplying facilities for project leaders to test their ideas on the public domain of Paris (land, buildings etc.) for 3 years, and in helping them in their funding search.

At the beginning of July 2014, NatureParif (the regional service for biodiversity promotion and nature conservation) and numerous partners involved in UA organized the first Summer Workshop on Urban Agriculture, attended by researchers, institutions and UA operators, including all the elected project leaders of Végétalisations innovantes. At regional level, a Call for Projects titled Filières agricoles de proximité (proximity agricultural supply chains) aims to provide financial support for some projects, mainly concerning the organization of peri-urban producers, but in some cases also new intra-urban production, to meet urban dwellers' growing demand for local products.

Last but not least, the new elected mayor of Paris, Anne Hidalgo, claimed during her spring 2014 campaign that she would aim for 100 ha of green rooftops in Paris, including 30% in diverse forms of urban agriculture, until the end of her mandate in 2020. Some months later, various services of the Mayor of Paris have been mobilized to concretize this objective and are working directly with our research team to exchange data and expertise about "how to do it". The "ParisCulteurs" Project will led to the setting up of UA projects in 2017 and 2018, with two successive calls.

This multiform political support is obviously of importance for accompanying the development of innovative forms of urban agriculture in Paris. Nevertheless, the project leaders are also facing many challenges. Some major obstacles were underlined during the meetings we organized with them. Three main concerns are:

- The real accessibility of space (on the ground or chiefly on a building) is very complicated, even when land or buildings belonging to the City of Paris are concerned. Technical data are not always available where necessary, additional work to reinforce rooftops for example or to increase their accessibility can be very expensive, and their funding is generally unclear. Legal rights to set up a farm on a roof are particularly difficult to obtain, especially regarding possible damages to the building and/or public accessibility. Project leaders consequently complain that they have lost a lot of time to motivate, to inform and to prove the feasibility of their projects.

- If they want to sell agricultural products directly, the project leaders would normally have the legal status of a farmer; but the "urban farmer status" does not exist in France and the Ministry of Agriculture is just beginning to realize that these new forms of agriculture may need some regulations.
- There are important disparities among the products' sale-price for the consumers. Among the urban soil and off ground projects (types 1–2; 14 existing sites), we identified four projects leaders' attitudes:
    - three projects directly address "top-of the range" customers, for example through sales or deliveries to gastronomic restaurants;
    - six have aligned their prices with organic market value, offering their products in specialized markets, boutiques or through delivery services, e.g. community supported agriculture (CSA);
    - two of them target families in difficulty, notably through solidarity groceries;
    - the remaining three harvest their products for personal consumption, but offer mainly other services (e.g., team building, cultural events, educational activities).
- Edible produce inside the city is automatically suspected of pollution risks, but measured data on the topic are scarce and sometimes controversial. For example, in collective gardens in Berlin, it was found that a significant percentage of the vegetables had high levels of heavy metals due to soil pollution and/or to traffic proximity (Säumel et al. 2012). On the other hand, in the AgroParisTech rooftop garden, measured levels of heavy metals in lettuce and tomatoes were much lower than European norms every year (Grard et al. 2013). It is nevertheless obvious that future urban producers will have to provide evidence of the safety of their products in terms, at least, of heavy metal content, which is regulated by European norms. Most of the project leaders consider this point as a form of discrimination compared to their peri-urban colleagues, who are not subjected to such an obligation, even if controls can always be carried out at various points of the supply chain. Recent local studies also show that traffic pollutant deposits on peri-urban plots may exist in some cases, as some of these market garden farms are deliberately set up near the communication routes so as to facilitate access to the Rungis' Wholesale Market (Petit et al. 2011). But in the case of intra-urban products, it is very probable that "future farmers" may be obliged to analyse their products frequently, to reassure consumers. One of the main questions is then: how to cover the costs of such tests, and how could the Paris authorities help these new producers in this respect?

9.5 Conclusion

Even though they have not yet defined a real food supply strategy for the City or the region, as some northern American cities have (e.g. Toronto with the GrowTo initiative (GrowTO 2012, and food policy councils in Seattle, Chicago etc. (Mansfield and Mendes 2012), the Parisian authorities are supporting various forms of UA in different ways, whether for their contribution to local short supply chains (a major topic for Paris and Region) or for other functions, including social and educational ones (recreational, social, educational, etc.).

For research, these innovative projects raise numerous questions, from the technical design of the cropping systems to their insertion, competition and complementarities with other forms of short supply chains. This latter point has for the moment not been properly addressed due to the novelty of these intra-urban projects. They also raise questions on functions other than the food supply they could represent for the city, for instance in terms of biodiversity enhancement and/or their contribution to the reduction of heat islets, and of runoff, and the absorption of some of the urban waste. Their comparison with "simple" (i.e. not productive) green rooftops or walls are important: the available results on the environmental effects of these walls are themselves widely diverse (Mentens et al. 2006; Simmons et al. 2008; Madre et al. 2014). We therefore imagine that productive forms of green rooftops could have even more heterogeneous effects. And even if all these environmental questions could be a centre of cities' interest in urban agriculture in the near future, the social adoption of these forms of agriculture must not be underestimated: it is probably largely in relation to their social acceptability that these innovative forms of urban agriculture will be able to develop in the next few years in Paris and Region (so for which types of urban dwellers do they cater, what are their relationships with the local inhabitants, and the benefits that these inhabitants can derive from the presence of a UA project on "their" buildings, for example?).

References

Abad R (2002) Le Grand Marché: l'approvisionnement de Paris sous l'Ancien Régime. Fayard, Paris

Agreste (2013) Infos rapides. Agreste Infos Rapides. Available via http://www.agreste.agriculture.gouv.fr/IMG/pdf/conjinfoleg201311toma.pdf

APUR (2013) Étude sur le potentiel de végétalisation des toitures terrasses à Paris, 39p. Avaiblable via http://www.apur.org/sites/default/files/documents/vegetalisation_toitures_terrasses.pdf

Aubry C, Chiffoleau Y (2009) Le développement des circuits courts et l'agriculture périurbaine: histoire, évolution en cours et questions actuelles. Innov Agronomiques 5:41–51

Aubry C, Kebir L (2013) Shortening food supply chains: a way for maintaining agriculture close to urban areas? the case of the French metropolitan areas of Paris. Food Policy 41:85–93

Barles S (2009) Urban metabolism of Paris and its region. J Ind Ecol 13(6):898–913. https://doi.org/10.1111/j.1530-9290.2009.00169.x

Brunet P, Charvet JP (coord) (1994) Le devenir des agricultures périurbaines. Bulletin de l'Association Française des Géographes Français 71(2):117–126

Cohen N, Reynolds K, Sanghvi R (2012) Five borough farm. Seeding the future of urban agriculture in New York City. A design trust for public space-added value. Chou Editor, New York

Coleman E, Petit E (2013) Des légumes en hiver, produire en abondance, même sous la neige. Actes Sud, Paris

Daniel AC (2013) Rapport sur l'agriculture urbaine et ses formes innovantes en France. AgroParistech, Paris

Daniel AC, Aubry C, Devins A (2013) Naissance et développement des formes commerciales d'agriculture urbaine en région parisienne. In: Agriculture Urbaine: Aménager et nourrir la ville? Vertigo, p 389

Grard B et al (2013) Rooftop and pollution: assessment of trace metals contamination in vegetables and in growing substrates made of recycled organic waste for an experiment on a rooftop in Paris. France. Poster at the urban environmental pollution symposium, Beijing, China, November

Grard B et al (2015) Recycling urban waste as possible use for rooftop vegetable garden. In future of food. J Food Agric Soc 3(1):21–34

GrowTO (2012) An urban agricultural action plan for Toronto (GrowTO action plan). Toronto, Toronto Public Health. Available at: http://tfpc.to/wordpress/wp-content/uploads/2012/08/GrowTO_ActionPlan_Oct161.pdf

Levidow L, Price B, Psarikidou K, Szerszynski B, Wallace H (2010) Urban agriculture as community engagement in Manchester. Urban Agric Mag 24:43–45

Madre F, Vergnes A, Machon N, Clergeau P (2014) Green roofs as habitats for wild plant species in urban landscapes: first insights from a large-scale sampling. Landsc Urban Plan 122:100–107

Mansfield B, Mendes W (2012) Municipal food strategies and integrated approaches to urban agriculture: exploring three cases from the global north. Int Plan Stud 18(1):37–60. https://doi.org/10.1080/13563475.2013.750942.

McClintock N, Cooper J, Khandeshi S (2013) Assessing the potential contribution of vacant land to urban vegetable production and consumption in Oakland, California. Landsc Urban Plan 111:46–58

Mentens J, Raes D, Hermy M (2006) Green roofs as a tool for solving the rainwater runoff problem in the urbanized 21st century? Landsc Urban Plan 77:217–226

Moreau JG, Daverne JJ (1845) Manuel pratique de la culture maraîchère de Paris. Librairie de la société, Paris

Morgan K (2009) Feeding the city. The challenge of the urban food planning (special issue). Int Plan Stud 14(4):341–348

Morgan K, Marsden T, Murdoch J (2006) Networks, conventions and regions: theorizing « worlds of food ». In: Place, power and provenance in the food chain. Oxford University Press, Oxford/New York, pp 7–25

Paddeu F (2012) Faire face à la crise économique à Détroit: les pratiques alternatives au service d'une résilience urbaine. Annales de Géographie 4:199–239

Paddeu F (2015) L'agriculture urbaine à Détroit: un enjeu de production alimentaire en temps de crise ? POUR 224:89–99

Petit C, Aubry C, Rémy E (2011) Agriculture and proximity to roads: how should farmers and retailers adapt? Land Use Policy 28:867–876

Phlipponeau M (1956) La vie rurale de la banlieue parisienne étude de géographie Humaine. Armand Colin, Paris

Porter JR, Dyball R, Dumaresq D, Deutsch L, Matsuda H (2013) Feeding capitals: urban food security and self-provisioning in Canberra, Copenhagen and Tokyo. Global Food Security https://doi.org/10.1016/j.gfs.2013.09.001

Poulot M (coord) (2013) Agriculture dans la ville, agriculture pour la ville: acteurs, pratiques et enjeux. Bulletin de l'Association Française des Géographes 90(3):276–285

Poulot M (2015) Agriculture et ville: des relations spatiales et fonctionnelles en réaménagement. Une approche diachronique. POUR 224:51–66

Poulot-Moreau M, Rouyres T (2000) La ceinture maraîchère et horticole francilienne, entre production économique et production de paysage. Méditerranée 95(3–4):51–57

Pourias J (2014) Production alimentaire et pratiques culturales en agriculture urbaine. Analyse agronomique de la fonction alimentaire des jardins associatifs urbains à Paris et Montréal. Thèse AgroParisTech, Université du Québec à Montréal, 249 p + annexes

Pourias J, Duchemin E, Aubry C (2015a) Products from urban collective gardens: food for thought or for consumption? Insights from Paris and Montreal. J Agric Food Syst Commun Dev 5(2):175–199

Pourias J, Aubry C, Duchemin E (2015b) Is food a motivation for urban gardeners? Multifonctionality and the relative importance of the food function in urban collective gardens of Paris and Montréal. Agric HumValues 33(2):257–273

Quellier F (2003) Des fruits et des hommes: l'arboriculture fruitière en Ile-de-France. (Vers 1600–vers 1900). Presses Universitaires de Rennes, Rennes

Redlinghöfer B, Traversac JB, Messmer JG, Aubry C (2014) Multi-actor organization for urban food systems: short but collaborative supply chains. Paper presented at the IFSA conference, Berlin, 1–4 April

Roy JM (2012) Les marchés alimentaires parisiens du XVII au XIXème siècles. Histoire Economie et Sociétés 17(4)

Sanyé-Mengual E, Ceron-Palma I, Oliver-Sol J, Montero JI, Rieradevall J (2012) Environmental analysis of the logistics of agricultural products from rooftop greenhouses in Mediterranean urban areas. J Sci Food Agric 93:100–109

Säumel I et al (2012) How healthy is urban horticulture in high traffic areas? Trace metal concentrations in vegetable crops from planting within inner city neighbourhoods in berlin, Germany. Environ Pollut 165:124–132. https://doi.org/10.1016/j.envpol.2012.02.019

Simmons M, Gardiner B, Windhager S, Tinsley J (2008) Green roofs are not created equal: the hydrologic and thermal performance of six different extensive green roofs and reflective and non-reflective roofs in a sub-tropical climate. Urban Ecosyst 11:339–348

Specht K et al (2013) Urban agriculture of the future: an overview of sustainability aspects of food production in and on buildings. Agric Hum Values 31(1):33–51. https://doi.org/10.1007/s10460-013-9448-4

Thomaier S et al (2014) Farming in and on urban buildings: present practice and specific novelties of zero-acreage farming (ZFarming). Renew Agric Food Syst 30:1–12. https://doi.org/10.1017/S1742170514000143

Toulallan M (2012) Les enjeux de l'approvisionnement alimentaire de la région ile-de France, rapport du conseil économique et social régional (CESR) au nom de la Commission de l'agriculture, de l'environnement et de la ruralité, Paris

Zasada I (2011) Multifunctional peri-urban agriculture – a review of societal demands and the provision of goods and services by farming. Land Use Policy 28(4):639–648. https://doi.org/10.1016/j.landusepol.2011.01.008

Chapter 10
Zero-Acreage Farming: Challenges and Opportunities for Urban Policies and Partnerships

Susanne Thomaier

Abstract In recent years a growing number of urban farming projects have been established in and on buildings. The term "Zero-Acreage Farming" (ZFarming) describes the idea of growing food without using any additional land or acreage. It encompasses open-air rooftop farms, rooftop greenhouses, productive facades, and indoor farming on and in existing or newly built urban structures. Using the urban building stock instead of farmland or vacant parcels involves very distinct opportunities and challenges: specific building-related challenges, technical restraints, regulatory frameworks, reluctance of landlords and developers, but also opportunities for enhanced resource efficiency and the creation of new urban spaces. The objective of this contribution is to examine these peculiarities and discuss the specific opportunities and challenges they imply. The study illustrates site-related requirements, legislative frameworks as well as specific economic risks and opportunities; gives an overview of different stakeholders involved in planning and implementation processes, stressing the role of landlords and developers – their expectations, motivations and fears; and deduces implications for policies, programs and stakeholder management.

Keywords Rooftop farming · Indoor farming · Planning challenges · Stakeholders

10.1 Zero-Acreage Farming: Urban Farming in a New Dimension

While land in many cities is scarce and expensive, the urban building stock offers a multitude of unused spaces – on rooftops but also in vacant buildings. Berlin, for example, has about 7302 available rooftop spaces larger than 500 m^2, totalling up to

---

S. Thomaier (✉)
Institute of Urban and Regional Planning, Technische Universität Berlin, Berlin, Germany
e-mail: s.thomaier@isr.tu-berlin.de

C.-T. Soulard et al. (eds.), *Toward Sustainable Relations Between Agriculture and the City*, Urban Agriculture, https://doi.org/10.1007/978-3-319-71037-2_10

8,317,935 m^2. that could be used for commercial rooftop farming (ZFarm 2015). Furthermore, there is a considerable number of very large and vacant buildings in Berlin for which "traditional" tenants are hard to find given their specific qualities. A study for the City of Toronto estimates that approximately 50 million m^2 of roof area are available for green roofs (Banting et al. 2004).[1]

Consequently, a growing number of urban farming projects have been established as rooftop or indoor farms in recent years using these untapped spaces. The term "Zero-Acreage Farming" (ZFarming) describes the very specific idea of growing food without using any additional land or acreage (Specht et al. 2013; Thomaier et al. 2014). It encompasses open-air rooftop farms, rooftop greenhouses, productive facades, and indoor farming on and in existing or newly built urban structures.

The farming techniques applied by ZFarming projects are soil-based or hydroponic and may involve livestock. The conceptual designs range from small-scale open-air container-based farming to large-scale rooftop greenhouses or indoor farms. ZFarms are either integrated as additional uses into buildings or established in buildings entirely dedicated to use by farming/food businesses. Most of the latter are located in formerly vacant buildings (e.g. warehouses) that are reused or redeveloped by indoor farms. The goals of ZFarming are quite similar to ground-based urban farming, including recreation, community building, environmental issues, education and awareness building, commercial activities and the creation of new agrifood business models (Thomaier et al. 2014, Cohen and Reynolds 2014). In developed countries, a striking difference from ground-based urban farming is the increasing number of commercial ZFarming businesses that operate controlled-environment rooftop greenhouses and indoor farms. Especially for this type of commercial projects, using the urban building stock instead of farmland or vacant parcels involves very distinct opportunities and challenges: specific building-related challenges, technical restraints, regulatory frameworks, reluctance of landlords and developers, but also opportunities for enhanced resource efficiency and the creation of new urban spaces. Because of this distinctiveness ZFarming adds some new dimensions to stakeholder management and policy making in urban agriculture. Paul Hardej, founder of "farmed here", an indoor farming business in the U.S., succinctly summarizes this by saying that vertical indoor farming is "land planning, designing, farming, building, manufacturing, food processing, distributing, marketing, selling, educating, researching, solving urban social problems" all in one, and that "governments don't know what to do with it (...)" (Hardej 2014).

[1]The criteria relied on by different studies on available rooftop spaces for green roofs or rooftop farming vary substantially. Therefore, numbers cannot be compared. For Berlin, the calculation is based on the available area, on flat rooftops, for solar energy. Besides a minimum size of 500 m^2, further criteria (building height, use and urban form of the building block) were taken into consideration. Static and structural features of the buildings were not included (ZFarm 2015). For Toronto, "[t]he benefits on a city-wide basis were calculated based on the assumption that 100% of available green roof area can be used. The available green roof area included flat roofs on buildings with more than 350 sq. m. of roof area, and assuming at least 75% of the roof area would be greened. The total available green roof area city-wide was determined to be 5000 hectares (50 million sq. m.)" (Banting et al. 2004).

The objective of this contribution is to examine these peculiarities and discuss the specific opportunities and challenges they imply for policy-makers and stakeholders.

Therefore I will illustrate site-related requirements, legislative frameworks as well as specific economic risks and opportunities from the perspective of ZFarming operators. Then, I give an overview of different stakeholders involved in implementation and operating processes, stressing the role of landlords and developers – their expectations, motivations and fears. Finally, I deduce implications for policies, programs and stakeholder management.

10.2 Methodology and Empirical Basis

This contribution is based on research conducted as part of the project "Zero-Acreage Farming" that was funded by the Federal Ministry of Education and Research and involved three German research institutions.[2] A qualitative approach was used, drawing on multiple methods. The findings presented here are derived from the following empirical basis:

Analysis of Existing ZFarming Projects A total of 76 projects in North America (44), Europe (19), Asia (15), and Australia (1) were identified by desk research in 2011 and 2012.[3] Given the project's focus on the role of ZFarming in developed countries, research was limited to Europe, North America and parts of Asia.[4]

The identified projects were analysed according to the following criteria: strategic orientation of ZFarming projects; spaces used; farming methods; ZFarming products and activities; market orientation; financing; and planning and implementation processes. Figure 10.1 illustrates the diversity of the investigated ZFarming projects based on selected criteria.

In-Depth Interviews In order to specify the desk-research, seven in-depth interviews with pioneers in rooftop farming in New York City were conducted. Interviewees were chosen from the following groups:

- ZFarming projects/operators
- Real-estate sector
- Planning and construction

[2] Technische Universität Berlin; inter3 – Institut für Ressourcenmanagement GmbH; ZALF – Leibniz-Zentrum für Agrarlandschaftsforschung e.V.

[3] This list is not supposed to be complete or statistically representative. It rather serves the purpose of providing a broad overview of ZFarming practices. Projects that have been realised after 2012 have not been part of the original analysis, but their development has also been studied and relevant findings have been considered for this contribution.

[4] i.e. Asian economies with upper-middle-income levels (according to World Bank definitions).

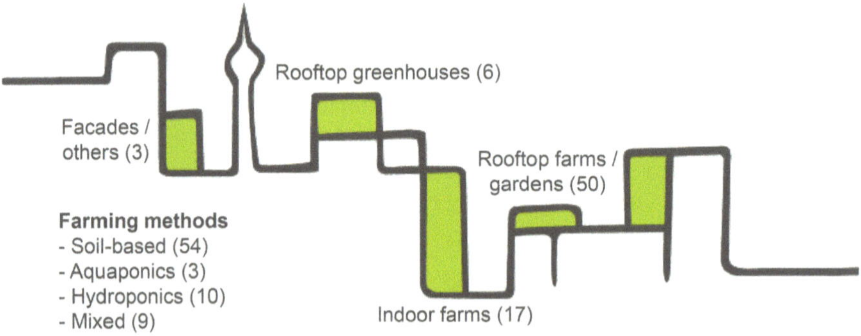

Fig. 10.1 Characteristics of the analysed ZFarming projects (n = 76)

Workshops with Potential Stakeholders in Berlin During the research project, five stakeholder workshops were held in Berlin between 2011 and 2013. The objective was to work out the potential benefits, opportunities and challenges associated with the implementation of ZFarming projects in Berlin. The themes discussed covered the entire process from preliminary planning to ZFarm operation (Specht et al. 2015).

10.3 Requirements for Urban ZFarming

When planning a rooftop or indoor farm, numerous requirements must be considered and examined. They largely depend on the project's objectives and conceptual layout as ZFarm designs range from low-tech to high-tech, from soil-based container farming to sophisticated controlled-environment farming. The most basic and relevant criterion in this respect is whether a project is supposed to be commercial or non-commercial and how its operational activities will be organized. Key issues and challenges are

- design-related,
- constructional,
- related to legislations and regulations and
- related to specific economic risks and opportunities.

10.3.1 Design-Related Criteria

Commercial ZFarms require large and, in the ideal case, homogenous spaces for large-scale production, whereas non-commercial projects can also be installed in smaller and more contorted spaces (Fig. 10.2).

Fig. 10.2 Commercial and non-commercial rooftop farms. Top left: Brooklyn Grange (author's picture). Top right: The Lufa Farms' greenhouse (Design + Environment, www.flickr.com, CC-BY 2.0). Bottom left: Schaduf rooftop garden (Anna Galda). Bottom right: Graze the Roof (Sergio Ruiz for SPUR, www.flickr.com, CC-BY 2.0)

While for indoor farming, artificial lighting is an indispensable prerequisite, sun exposure and shading of a roof are pivotal criteria when choosing a site for rooftop farms (open-air and greenhouses).

When designing rooftop farms, potential nuisances for the neighbourhood – e.g. by light reflections from the greenhouse or by artificial greenhouse lighting – have to be considered. Hence, the position of adjacent buildings plays a relevant role when choosing a site.

ZFarming must neither interfere with the immediate environment nor with other uses within the building. Ideally, other tenants may even benefit from the ZFarm as an amenity. Accordingly, accessibility issues have to be considered. Accessibility depends on the conceptual layout of a ZFarm project, e.g. whether or not it is open to the public. Furthermore, commercial ZFarm operations have to accommodate on-site logistics, i.e. spaces for packaging, parking and loading, which may imply additional space requirements and redevelopment costs.

10.3.2 Constructional Criteria

Challenges related to the integration of ZFarms into the existing building stock depend on the conceptual layout and type of each ZFarm.

In any case, a building's weight load capacities (and in the case of rooftop farms its snow and wind load resistance) are decisive for ZFarming. Of course structural reinforcements can be undertaken, subject to the owner's financial resources and consent and relating to the general condition of the building. Depending on the age and prior use of a vacant building, possible contamination and food safety have to be considered, especially if a food business is to be established.

Interviews have shown that chances for adding a ZFarm structure to an existing building rise when there is a general need for retrofitting or renewing the building as construction work will be undertaken, anyway. Generally, infrastructural issues play an important role, for example mechanical equipment like vents, or water and energy infrastructure influence the design and feasibility of ZFarms (Sanyé-Mengual et al. 2014). Various studies prove the potential of ZFarming to raise a building's energy efficiency by improving its isolation or by enabling thermal synergies between rooftop greenhouse and building (Castleton et al. 2010; Liu and Minor 2005; Delor 2011; Pfoser et al. 2013). Often, however, these potential advantages cannot be fully exploited, e.g. when the existing building infrastructure is not suitable for connecting energy and water cycles of the farm and the building or if realizing them would involve high costs.

> (They) didn't want us to change the building too much. We looked at putting a heat pump (…) but it was not allowed. (…) The building has a steam system, but (…) it was not reliable (…) We also looked at using gas (…), (what) was too complicated. So we ended up with an electric heat pump. (Interviewee 2; Planning and construction)

Also the time lines for the (re)development of a building and a ZFarm in or on it may differ:

> One of the tricky things is that they are redeveloping the whole building. So they don't know yet who the tenants are and where to put the mechanical equipment on the roof. So we can't design our greenhouse until we know where everything is. (Interviewee 1; Planning and construction, ZFarming projects/operators)

10.3.3 Land Use Regulations and Building Codes

The eligibility of and permission for a ZFarm project at a chosen site are dependent on legislative frameworks of the public building law. Paul Hardej calls the process of getting entitlements the actual plowing, and the process of complying with building codes the actual seeding of urban (indoor) farming (Hardej 2014). Much of our interview information concerning these issues relates to the U.S. context, while the background for structuring the selected key criteria mentioned below relates to German legislation. Even though the legislative frameworks may substantially differ between the two countries, interviewees in the U.S. as well as Germany stress the issues outlined below as decisive in both countries. Things may be regulated on different levels and in terms of different legislations, but the regulative issues and challenges are the same. They mainly relate to zoning, building codes and a range of ancillary regulations.

Depending on the respective ZFarming concept, a ZFarm may add a new building use (e.g. commercial farming on housing or in a manufacturing building) or change the use that has to conform to zoning regulations. This mostly concerns commercial ZFarming because agricultural businesses (with or without livestock) are not allowed in all land use zones. Some municipalities do not have zoning codes that allow farming, even though many cities have already adapted their zoning codes accordingly. Zoning ordinances might also differentiate between production/growing, processing and sale. This means additional challenges for commercial ZFarming projects that integrate all these activities.

Regulations and limitations concerning the building height, number of floors, and floor-area-ratio may pose restrictions on rooftop greenhouses and open rooftop farms. Many existing buildings have often already reached their maximum for the respective criteria. In some cities rooftop greenhouses are rated as a full additional floor (depending on building codes and regulations), thus ruling out their implementation on many buildings. If a ZFarming structure is from the very start integrated into the architectural layout of a new building, architects can of course design everything in compliance with regulations.

Besides, a broad range of further bylaws, building regulations and codes may influence the feasibility of ZFarming at a specific site. Critical issues are: distance space to the neighbouring building, fire safety regulations, energy issues, wind load resistance, load capacities and parapets (in case of rooftop farms or gardens). These legal regulations may require an increase in the materials needed to comply with various safety standards, which influences a project's ecological and economic feasibility (Sanyé-Mengual et al. 2015a).

In addition to aspects that are inherently connected to constructional issues, ancillary regulations, which apply to potential external impacts of ZFarms, have to be considered. These may include emission protection regarding noise, odour, light or glare effects (e.g. in the case of greenhouses) as well as urban heritage conservation or waste regulations.

The above gives an idea of the variety of regulations that have a bearing on ZFarming and shows that complying with existing regulations and policies is fraught with challenges. In some cases, experiences with the first built rooftop farms have acted as a driving force for adapting regulations and policies to the innovative nature of ZFarming. Cities like New York and Boston amended zoning resolutions to encourage educational and commercial farming, including rooftop food production (Cohen et al. 2012) (City of Boston 2014). The New York resolution allows for greenhouses complying with certain regulations to be exempt from height restrictions and floor area limitations. Chicago included (commercial) rooftop farming into its zoning code (City of Chicago 2014). However, such adjustments are still at a early stage and cannot be extrapolated to other national settings with different planning and regulatory systems.

In their study on the acceptance of urban rooftop farming (URF) in Barcelona, Sanyé-Mengual et al. (2015b) state that URF legitimacy could be raised by a greater endorsement of new projects by various municipal departments. "These departments may play key roles in the revision of the legislation, in the development of local policies to promote local production, and in the dissemination of information on the benefits of URF" (Sanyé-Mengual et al. 2015b).

10.3.4 Specific Economic Risks and Opportunities

The economic feasibility of ZFarms strongly depends on their conceptual layout. Especially enclosed ZFarms such as rooftop greenhouses or indoor farms might involve technical and constructional obstacles with high initial costs for retrofitting the space and constructing the farm with all its technical equipment (Banerjee and Adenäuer 2014; Sanyé-Mengual et al. 2015a).

In terms of economic feasibility, all organizational, procedural, temporal, personnel-related, structural and infrastructural aspects need to be taken into account. Possible monetary expenses include not only investment, financing and running cost, but also consultancy and transaction cost, which might be unexpectedly high. Due to the high degree of innovation, planning and implementation processes are complex, and there still is a lack of ZFarming experience in many places – also concerning the eligibility for grants – both on the administrative side and on the side of stakeholders and experts.

> The building department was tricky (…), but just because they didn't know how to consider (the project). (Interviewee 2; Planning and construction)
>
> I met with them in order to see if we could help to finance the project and it was like I was speaking Greek to them. So there is a huge learning curve (…) We were in these early stages of solar, now we are in these early stages of (…) rooftop agriculture particularly (…). (Interviewee 3; Real estate sector)

As a consequence, the time and money needed to convince investors and other stakeholders, work out a detailed conceptual ZFarm design and get building and business permissions often exceed expectations. However, planning processes tend to accelerate due to learning processes and knowledge building, as Lufa Farms founder Mohamed Hage confirms: "It took me 3–4 years initially for my projects. Now it's about 1-1, 5 years from start to finish. It's still a challenge, and we need to come down to under 1 year for it to make sense economically" (Hage 2013). Establishing close collaborations that brings together all the different expertise is crucial for reducing time requirements and costs.

Apart from economic risks associated with planning issues, commercial ZFarming also faces challenges concerning the food market and potential sales. Owing to their small size and, accordingly, small yields and higher unit costs as well as their explicit focus on local markets, typical distribution channels, including wholesale, are often not appropriate for commercial farms. Therefore, alternative supply chains without intermediaries but strong reliance personal contacts and collaborations have to be established. The need for direct distribution with no or very low trade margins encourages alternative, collaborative funding and distribution models (e.g. CSA) and the organization of new local food collaborations. Lufa Farms, for example, is partnering with other local food producers to fill out product offerings (e.g. bread, cheese, honey). Many commercial ZFarms operate along the entire supply chain and additionally expand their activities beyond mere food production (e.g. social activities, events, education), thus fostering local community networks.

Studies have shown that customers are sceptical towards soilless growing methods such as hydroponics and aquaponics (Specht et al. 2016; Sanyé-Mengual et al. 2015b), which are often applied in enclosed commercial ZFarms. Hence, customers' acceptance can be an economic risk. On the other hand, using efficient growing methods in enclosed environments allows for year-round production with constant, high yields and substantially reduced risks for crop failure. This may induce retailers to become partner businesses.

The innovative nature of ZFarming involves further opportunities for marketing. Its (initial) uniqueness constitutes a valuable benefit in terms of image and brand creation.

Especially the notion of fresh local produce throughout the year is interesting for tapping consumer markets.

> The goal was also to have tomatoes at times when you don't get good ones from others. (...) the customers definitely appreciate the locally grown food (...). (Interviewee 5; ZFarming project/operator)

These examples show that commercial ZFarmers not only need comprehensive knowledge of farming methods that fit into built environments but may also need marketing and distribution competences.

10.4 New Partnerships for Urban ZFarming

10.4.1 Multi-stakeholder Partnerships

Due to the peculiarities of the planning and implementation process associated with ZFarming, it is worth taking a closer look at the involved stakeholders, regimes and their interactions. Karge (2015) did a comprehensive stakeholder analysis of the non-profit rooftop garden project himmelbeet in Berlin. The open-air rooftop garden was designed as a community garden with raised beds, a small café, and a space for small events. Karge (2015) identified a plurality of influencing factors and players in the following three fields:

- institutional regimes: legislations, regulations, grants, foundations;
- economic and public supporters: architects, urban gardening consultants, various supporting companies (green roof company, sponsors, retailer, main tenant of the building, lawyer), and the municipal administration;
- individual players: core team and associates, volunteers, gardeners, general supporters.

Hence, stakeholders and experts from various fields are involved in different stages of the planning process. Depending on the respective project, they contribute with ideas/knowledge, networking, financing, workforce, or power.

The interplay of developing the idea, deciding on a site and organizing the capital input, differs between projects. And so does the contribution of the partners to a project. In the early stages of ZFarming – and this is probably still the most common process – ZFarms were initiated by a core team of the farming business or initiative with the idea of a farm in mind, looking for a suitable site and financing. Once ZFarms proof to be a successful model, projects also happen to be initiated by landlords or developers, approaching ZFarming operators.

> So (the Department of Environmental Protection) did a competition to see who could come up with the best projects. And that's when we reached out to (them), who already had the farm (…). So they already had a model that worked, they had done it and they wanted to expand, so we teamed up to apply for the grant. (…) their job is not only to run their business (…) but also to maintain the roof for us. So it's such a Win-Win (…). (Interviewee 4; Real estate sector)

Accordingly, the capital input of the involved stakeholders differs enormously. Landlords and developers might invest a certain amount of money to get the required infrastructure set up on the rooftop, while the construction of the farm itself with all the associated planning cost is financed by the ZFarm initiative.

> So we had to spend – more time than money – but it also took money, to move some of the infrastructure so that we could give (them) kind of a footprint of the roof, that was big enough for them. (…) we agreed to spend up to X $. And anything above X they had to pay. And with that in mind we moved some vents and we also reconfigured and upgraded the existing solar array (…). (Interviewee 3; Real estate sector)
>
> So we knew in order to apply for the grant we had to ask for a certain amount of money (…) We had a landscape architect who set up a budget. So we got (the greatest part) from the city, we put (something) in and they (the farmers) put (something) in. (Interviewee 4; Real estate sector)

However, such an (financial) involvement of the building owner is not always given. Especially smaller projects might carry out the planning, financing and implementation largely on their own, with external consultancy and/or voluntary support.

A comparison of Gotham Greens' first commercial greenhouse project in New York and the community-based project himmelbeet in Berlin show some essential differences concerning involved stakeholders and their management (Table 10.1). Gotham Greens as well as himmelbeet are organized as limited liability companies, with Gotham Greens being for-profit and himmelbeet being non-profit.

Both projects faced problems in finding a suitable site, mainly due to fire safety and other planning regulations (Karge 2015 and Gotham Greens LLC 2011). While Gotham Greens could find an alternative rooftop space after a first failed try, himmelbeet had to fall back on a ground-based site. A facilitative factor for himmelbeet in finding this site was its well prepared planning and its local embeddedness with close networks. The good networking of the himmelbeet team with local politics

Table 10.1 Involved players in planning and construction of Gotham Greens' computer controlled rooftop greenhouse and the community-based garden himmelbeet

		Gotham Greens	himmelbeet
Planning & development phase	Conceptional and system design	Architects	Architects
		Engineers	Landscape architect
		Commercial greenhouse manufacturer	Volunteers
			Urban gardening initiative as consultants
			Green roof specialist
		Construction management firm	Local business district for networking and financial support
		Gas and utility providers	Local businesses for financial support
			Foundation for conceptional and financial support as well as networking
			University and environmental association for knowledge support
	Construction approvals and permitting	Over 2 years interacting with several city departments and agencies	Less than 1,5 year permitting process from first idea to opening, including the rejection of the rooftop garden by the administration, the modification of the design to an alternative ground-based space and the final approval and opening of the garden
			Involved players:
			Architects
			Green roof specialists as consultants
			Main tenant of thc building
			City Departments and agencies
	Architecture, engineering, site preparation	Structural engineering firm	Architects
		Cost estimator	Volunteers
		Steel contractor	Main tenant of the building
		Structural engineer	Environmental associations
		Construction manager	
		Electrical, mechanical engineer	
		Testing agency to test for contamination	

(continued)

Table 10.1 (continued)

		Gotham Greens	himmelbeet
Implementation & construction phase	Construction	Material procurement:	Architects
		Greenhouse manufacturers	Landscape architect
		Equipment suppliers	Volunteers
		Hydroponic system providers	Patch renters/gardeners for workforce and financing
		Renewable energy installers	
		Horticultural supply vendors	Local businesses for financial support and workforce
		Construction management firm bided the construction of the facility to a variety of subcontractors	Public institutions for financing support
			Foundation for networking
			Association for delinquent persons for workforce
	Commissioning	Electrical engineers for computer climate control system	Environmental associations
		Food safety: food safety expert for reviewing and auditing the food safety plan	
	Sales and marketing	Whole foods market as one of the biggest customer helped with in-store marketing	Graphic designer for "corporate" design
			Patch renters
			Marketing on farmers markets and events
		Branding and design firm	On-site sale of produce
			On-site cafè

Own design according to Gotham Greens LLC (2011) and Karge (2015)

also helped to fasten the implementation process, despite the drawback of finding a new site and concept. The analysis of both cases suggests that the planning and construction process of Gotham Greens was much more professionalised with a construction management firm and a wide range of sub-contractors for various planning and construction duties. At himmelbeet all the activities and tasks have been managed by the core team and volunteers. Most partners came from a very local context and had rather supporting than business driven functions. Especially the numerous players involved in the system design and construction of the Gotham Greens greenhouse facility reflect the complexity of the stakeholder management. However, also the case of the community-oriented himmelbeet illustrates that implementing ZFarming often requires a higher degree of organization and institutionalization compared to ground-based urban farming since there are far more juridical and financial issues and risks involved. This was one reason why the himmelbeet team decided to set up a limited liability company at a very early stage of the planning process.

Germer et al. (2011) point out that the technical challenges of crop cultivation in buildings require interdisciplinary cooperation between experts who usually work independently in separate fields. High-tech commercial ZFarms, in particular, that face high challenges regarding permissions, food safety issues, design and construction, cannot do without experts from all disciplines, such as architects, agricultural engineers, food safety experts, environmental engineers and also marketing experts. Our interviews and workshops involved actors from most of the above-mentioned groups, attesting to the great need for interdisciplinary exchange and knowledge building.

Building owners and developers who support ZFarming also highlighted that integrated (design) processes are crucial, as the following quotations from landlords' statements show:

> There was a lot of work by a lot of different parties. (…) To coordinate that, then develop the plans, have the plans approved by the building department, then construct it and then have the constructed project approved by the building department (…) is a very laborious process. (Interviewee 3, Real estate sector)

The quote also documents that permitting processes of innovative projects such as rooftop greenhouses require endurance and resources.

What Heerwagen (2000) states for green buildings also holds for knowledge building in ZFarming: "Discussions with managers and members of the design team for a new green building (...) also suggest that technology transfer and learning may be a hidden benefit of sustainable design and construction, especially when techniques and technologies are new. If these benefits are accrued at the local level, then the transfer of skills to other building projects can benefit the community as a whole" (Heerwagen 2000).

This kind of learning process and knowledge transfer can also trigger policy making, which is described by a study on urban agriculture policy-making in New York: "(…) planning and policy making is also occurring in non-traditional spaces of interaction among practitioners, advocates, technical assistance providers, researchers, government officials, funders, and other individuals and organizations with a stake in the future of urban agriculture. These cross-sectoral networks and collaborative problem-solving processes come together at strategic moments, whether around a proposed local law, or an effort to collectively measure the productivity of gardens, and then may be reconfigured or disbanded" (Cohen and Reynolds 2014). Gotham Greens LLC (2011) confirm this in their report by stressing their participation in several hearings, workshops and meetings to help draft legislation and building code to support urban rooftop greenhouses.

10.4.2 The Real Estate Sector: A Crucial Partner for ZFarming

The support of the real estate sector is a decisive factor for ZFarming. The degree of diffusion of ZFarming as well as the motivation and reluctance of owners and developers vary greatly between countries. Our research revealed that in Berlin, where no large-scale ZFarm entities exist up to now, scepticism of real estate developers or landlords is still rather high whereas interviewees in New York, once committed to ZFarming, supported the projects by various procedural, financial or constructional means.

Heerwagen (2000) identifies two main strategies of proponents of green buildings, which also apply to owners and developers in a ZFarming context: they either expect an added value for the building or seek to reduce costs (Heerwagen 2000). As mentioned above, ZFarming raises a building's energy efficiency and contributes to rainwater retention (Carson et al. 2013), thus reducing energy and sewage cost. Landlords and developers are indeed aware of those resource- and cost-related benefits, but they are even more strongly oriented to adding value through ZFarming.

They perceive ZFarms as one means (sometimes in conjunction with others) to (re)develop a building in accordance with the principles of sustainable and "green" architecture. On the one hand, (particularly accessible open-air rooftop) ZFarms serve as attractive amenities for tenants offering recreational qualities and access to fresh food. On the other, ZFarms function as promotional features that attract (innovative) tenants and represent a certain image of the site and building. And even more, ZFarming businesses can be one means to an end for landlords when it comes to the maintenance of a green roof as they assume responsibility for it.

> They saw that the farm would help to make their building special and help make tenants want to go there. And they wanted to make it more sustainable. So for them it is also a marketing thing. (Interviewee 7; ZFarming project/operator)
>
> So we have got all the benefits of having the green roof but then have them running it as a business. They are up there, keeping it clean, making sure that it's always planted and putting cover crops in the winter. And then they sell in a market (…), and they supply the café here. (Interviewee 4; Real estate sector)

For most of the interviewed real estate stakeholders their commitment to ZFarming reflects a pioneering and open-minded attitude towards innovative and experimental ideas.

> (We have) often done things as somewhat of an experiment, almost like from a public policy perspective. (…) So we looked at this greenhouse operation as somewhat of an experiment also, and to see if we can maybe nurture it to something that could grow and be more successful elsewhere. (Interviewee 3; Real estate sector)
>
> (…) we wanted to do something innovative and the primacy was to design the next generation of green affordable housing. (Interviewee 6; Real estate sector)

This mindset is also one reason why pioneering ZFarm projects are often supported by non-profit (public) developers and landlords, who are rather willing to get involved in time consuming planning and implementation processes than purely profit-driven players.

So I don't think that they were able to do that with a kind of a regular landlord. Because we are a little bit different. First of all we are not profit. (...) I don't think that a typical landlord would have the patience to do something like this, because it was a long, long process. Even to the point when we negotiated the lease. (Interviewee 3; Real estate sector)

From the developers' perspective the cost and time investment is experienced as a great "leap of faith" since in many places, ZFarming is not yet an established business model and still awaits proof of its long-term economic feasibility. For ZFarming businesses, the set-up of a farm involves a large capital investment, why they need a high planning security.

That said, the drafting of leases plays an important role in terms of securities for both owners and ZFarming businesses as leases not only ensure long-term reliability and determine rent but may also define responsibilities for dismantling ZFarm structures. Compared to ground-based urban farming, commercial ZFarming businesses tend to have longer leases with durations from 10 to 30 years.

Normally if I would negotiate a lease with someone on a space, if that business failed, well I have the space and rent it and put someone else in there. (...) I kept saying: 'If you go out of business I have a greenhouse. So what am I supposed to do with that? (...) There is not another one of you and take along the greenhouse.' So that was kind of hard to them to get their heads around, because they were spending all that money to build this very fancy facility and I was saying: 'Well that's great, but it doesn't really do anything for us. Even though we are very kind of liberal minded in this process, we still have financial interest to protect.' (...) Basically money solved the problem. They have a larger security deposit with us. So if they do go out of business we are left with the greenhouse and we have to deconstruct it, but their security deposit will pay for some of the costs. (...). (Interviewee 3; Real estate sector)

Obviously, for open-air rooftop gardens or farms the problem of dismantling is easier to solve, since they can be easily re-used as "normal" green roofs and amenity spaces (Interviewee 4; Real estate sector).

Even though most owners generate income from rent from ZFarming businesses, this is – at least in the case of rooftop farms (indoor farms differ in this point) – not the main driving force for supporting them since farming businesses usually pay lower rents than other tenants. Rather, income from rent may be a decisive criterion for the owner's willingness to contribute financially to ZFarming-related infrastructure adjustments (e.g. for water or electricity) as rent offsets these investments.

ZFarms do not face land constraints like ground-based urban farming projects, but still they compete with other uses. If ZFarming is integrated into a new building, a developer or owner usually has to decide whether to exploit the allowed maximum intensity with a rooftop greenhouse or an additional floor of housing. Since financial returns will be higher from housing than from a rooftop greenhouse, this might be a hurdle for ZFarming. Also, solar energy generation is a favourable use for rooftops. While landlords may get financial incentives for installing roof-based solar energy systems, rooftop farms are not per se eligible for comparable incentives. Moreover, solar installations are clearly framed by proven policies and regulations, whereas innovative projects such as ZFarming involve high policies- and regulations-related insecurities. Generally, the availability of redevelopment capital (like in New York's manufacturing and industrial areas for example) facilitates the implementation and dissemination of urban agriculture and ZFarming (Ackermann et al. 2014).

10.5 Conclusion

The described peculiarities of ZFarming are mainly due to the use of urban real estate and the policies, regulations and actors involved. Hence, planning- and building-related challenges and opportunities are most decisive for ZFarming. The required technical and procedural innovations involve advancements in farming technologies, e.g. developing specific growing substrates and specific materials for rooftop greenhouses or optimizing energy-efficient LED-lights for indoor farming, on the one hand; and the need to adapt impeding policies, legislations and regulations in land use planning and, more specifically, zoning and building codes and regulations, as well as additional incentives for businesses to operate in the city, on the other.

Generating (new) expertise on constructional, technical and procedural issues helps to identify obstacles and to enhance knowledge building for both government-driven and stakeholder-led planning and policy-making processes. Collaboration and new partnerships among different stakeholders and consultancy help to provide information on ZFarming feasibility (see also Cohen and Reynolds 2014 and Sanyé-Mengual et al. 2015b). Practical guidelines for ZFarming initiators and municipal agencies enhance the dissemination of knowledge and streamline planning and implementation processes (see, for example, ZALF 2013). Integrating ZFarming into comprehensive plans or specific (municipal) programs increases public awareness, appreciates its potentials and spurs its diffusion, especially since it is largely consistent with superior goals and objectives of sustainable urban development. Important financial incentives for facilitating ZFarming are programs, grants and (tax) incentives in the following fields: green roofs, stormwater management, local economic and business development, rehabilitation of existing buildings, installation of renewable energy sources, closed resource cycles as well as green building certification programs like LEED.

Operational ZFarm business is also largely based on networks with social institutions, retailers, restaurants etc. Commercial ZFarms rely on networks with a variety of other food businesses for establishing local and direct food supply chains.

This contribution has shown how planning for ZFarming differs from familiar planning processes and what role different stakeholders play. In spite of challenges described, expanding businesses in Northern America show that roadblocks can be overcome and growing experiences spur dissemination. Timelines become shorter because expertise has been built, policies are slowly adapting and key players have been sensitised. However, regional differences have to be kept in mind – in terms of policies but also regarding the suitability of different ZFarming types for varying local contexts. Hence, the diffusion of different ZFarm types will most likely differ between countries, cities and even neighborhoods, because needs and local contexts are different. For example, we can see a rising number of indoor farms in Japan due to scarce space and food safety issues after Fukushima. In the U.S. rooftop farming, but also indoor farming is on the rise, whereas in Germany acceptance for technically sophisticated high-efficiency farming still seems to be rather low (Specht et al. 2015).

References

Ackermann K, Conrad M, Culligan P, Plunz R, Sutto MP, Whittinghill L (2014) Sustainable food systems for future cities: the potential of urban agriculture. Econ Soc Rev 45(2): 189–206

Banerjee C, Adenäuer L (2014) Up, up and away! The economics of vertical farming. J Agric Stud 2(1):40–60. https://doi.org/10.5296/jas.v2i1.4526

Banting D, Doshi H, Li J, Missios P, Au A, Curri BA, Verrati M (2004) Report on the environmental benefits and costs of green roof technology for the city of Toronto. http://www1.toronto.ca/city_of_toronto/city_planning/zoning__environment/files/pdf/fullreport103105.pdf. Accessed 2 Aug 2015

Carson TB, Marasco DE, Culligan PJ, McGillis WR (2013) Hydrological performance of extensive green roofs in New York City: observations and multi-year modelling of three full-scale systems. Environ Res Lett 8(2):1–13. https://doi.org/10.1088/1748-9326/8/2/024036.

Castleton HF, Stovin V, Beck SBM, Davison JB (2010) Green roofs; building energy savings and the potential for retrofit. Energy Build 42:1582–1591. https://doi.org/10.1016/j.enbuild.2010.05.004

City of Boston (2014) Article 89 made easy: urban agriculture zoning for the city of Boston. http://www.cityofboston.gov/images_documents/usersguide_july-1_tcm3-45895.pdf. Accessed 28 Feb 2015

City of Chicago (2014) Chicago zoning and land use ordinance. http://www.amlegal.com/nxt/gateway.dll/Illinois/chicagozoning/chicagozoningordinanceandlanduseordinanc?f=templates$fn=default.htm$3.0$vid=amlegal:chicagozoning_il. Accessed 28 Feb 2015

Cohen N, Reynolds K (2014) Urban agriculture policy making in New York's "new political spaces": strategizing for a participatory and representative system. J Plan Educ Res 34(2):221–234. https://doi.org/10.1177/0739456X14526453

Cohen N, Reynolds K, Sanghvi R (2012) Five borough farm: seeding the future of urban agriculture in new York City. Design Trust for Public Space, New York

Delor M (2011) Current state of building-integrated agriculture, its energy benefits and comparison with green roofs. University of Sheffield. http://e-futures.group.shef.ac.uk/publications/pdf/103_12%20Milan%20Delor%20summary.pdf. Accessed 2 Feb 2017

Germer J, Sauerborn J, Asch F, de Boer J, Schreiber J, Weber G, Müller J (2011) Skyfarming an ecological innovation to enhance global food security. J Consum Prot Food Saf 6(2):237–251. https://doi.org/10.1007/s00003-011-0691-6

Gotham Greens LLC (2011) Gotham Greens Farms LLC, sustainable urban CEA, final report for the New York State Energy Research and Development Authority. http://www.nyserda.ny.gov/Business-and-Industry/Agriculture/CEA-Completed-Projects. Accessed 2 Feb 2016

Hage M (2013) Contribution on the urban farming conference "Cultivating lands, nourishing communities, building businesses", Panel 5 rooftop techniques, Roxbury Community College, 9 March. https://citygrowers.wordpress.com/urban-farming-conference/panel-5-roof-top-techniques/. Accessed 28 Feb 2015

Hardej P (2014) Farmed here. Sustainable indoor farming. In: Presentation on the International conference on vertical farming and urban agriculture, University of Nottingham, 9–10 September. http://vfua.org/wp-content/uploads/2014/12/Paul-Hardej.pdf. Accessed 28 Feb 2015

Heerwagen J (2000) Green buildings, organizational success and occupant productivity. Build Res Inf 28(5/6):353–367. https://doi.org/10.1080/096132100418500

Karge T (2015) Neue Urbane Landwirtschaft – Eine theoretische Verortung und Akteursanalyse der Initiative Himmelbeet im Berliner Wedding (German). ISR Arbeitshefte 79. Technische Universität Berlin, Berlin

Liu K, Minor J (2005) Performance evaluation of an extensive green roof. Greening rooftops for sustainable communities. Washington, DC. http://www.sustainabletechnologies.ca/wp/wp-content/uploads/2013/03/NRC_EastviewGRrept.pdf. Accessed 2 Feb 2017

Pfoser N, Jenner N, Henrich J, Heusinger J, Weber S (2013) Gebäude, Begrünung und Energie: Potenziale und Wechselwirkungen. Interdisziplinärer Leitfaden als Planungshilfe zur Nutzung energetischer, klimatischer und gestalterischer Potenziale sowie zu den Wechselwirkungen von Gebäude, Bauwerksbegrünung und Gebäudeumfeld (German). Technische Universität Darmstadt, Fachbereich Architektur. https://www.irbnet.de/daten/baufo/20128035673/Kurzbericht.pdf. Accessed 2 Feb 2017

Sanyé-Mengual E, Llorach P, Sanjuan-Delmás D, Rieradevall J, Montero JI, Oliver-Solà J (2014) The rooftop greenhouse lab: optimising food production on buildings through integrated rooftop greenhouses in Barcelona, Spain. In: Presentation on the International conference on vertical farming and urban agriculture, University of Nottingham, 9–10 September. http://vfua.org/wp-content/uploads/2014/12/Pere-Llorach-Masana.pdf. Accessed 28 Feb 2015

Sanyé-Mengual E, Oliver-Solà J, Montero JI, Rieradevall J (2015a) An environmental and economic life cycle assessment of rooftop greenhouse (RTG) implementation in Barcelona, Spain. Assessing new forms of urban agriculture from the greenhouse structure to the final product level. Int J Life Cycle Assess 20:350–366. https://doi.org/10.1007/s11367-014-0836-9

Sanyé-Mengual E, Anguelovski I, Oliver-Solà J, Montero JI, Rieradevall J (2015b) Resolving differing stakeholder perceptions of urban rooftop farming in Mediterranean cities: promoting food production as a driver for innovative forms of urban agriculture. Agric Hum Values 33(1):101–120

Specht K, Siebert R, Hartmann I, Freisinger UB, Sawicka M, Werner A, Thomaier S, Henckel D, Walk H, Dierich A (2013) Urban agriculture of the future: an overview of sustainability aspects of food production in and on buildings. Agric Hum Values 31:33–51. https://doi.org/10.1007/s10460-013-9448-4

Specht K, Siebert R, Thomaier S, Freisinger UB, Sawicka M, Dierich A, Henckel D, Busse M (2015) Zero-acreage farming in the city of Berlin: an aggregated stakeholder perspective on potential benefits and challenges. Sustainability 7:4511–4523. https://doi.org/10.3390/su7044511

Specht K, Siebert R, Thomaier S (2016) Perception and acceptance of agricultural production in and on urban buildings (ZFarming): a qualitative study from Berlin, Germany. Agric Hum Values 33(4):753–769. https://doi:10.1007/s10460-015-9658-z

Thomaier S, Specht K, Henckel D, Dierich A, Siebert R, Freisinger UB, Sawicka M (2014) Farming in and on urban buildings: present practice and specific novelties of zero-acreage farming (ZFarming). Renew Agric Food Syst 30(1):43–54. https://doi.org/10.1017/S1742170514000143

ZALF Leibniz-Zentrum für Agrarlandschaftsforschung e.V (2013) There's something growing on the roof. Rooftop greenhouses – Idea, planning, implementation. Planning Guideline. Berlin. http://www.zalf.de/htmlsites/zfarm/Documents/leitfaden/Rooftop%20greenhouses.pdf. Accessed 28 Feb 2015

ZFarm Zero-Acreage Farming (2015) Potential ZFarming rooftop areas in Berlin. http://www.zfarm.de and http://kartographie.planen-bauen-umwelt.tu-berlin.de/bfarm/svg/index_svg.html. Accessed 2 Aug 2001

Chapter 11
The Emergence of Municipal Allotment Gardens in Greece in Times of Crisis. Governance Challenges for New Urban Gardening Practices

Theodosia Anthopoulou, Sofia Nikolaidou, Maria Partalidou, and Michael Petrou

Abstract Urban gardening has never been a tradition in Greece. Their recent growing number is mainly linked to the economic crisis affecting particularly urban households. The deepening crisis and the increasing urban (neo)poverty began to challenge local authorities to search for alternative ways of food (fresh and affordable) provisioning. Municipal Allotment Gardens, the prominent type of institutional response, are embraced by both the local authorities and citizens as alternative spaces within city neighborhoods for ensuring livelihoods and providing a way out of the multiple effects of the recent crisis. Drawing on empirical fieldwork from three different municipal allotment gardens, this study explores the institutional and political context of their establishment and local authorities' aspirations. In addition, through interviews and focus groups investigates motivations for applying for a municipal plot as well as lived experiences of gardeners shedding light on sustainability issues and future perspectives of these "crisis gardens". Either driven by the economic crisis, or other motivations such as community building, psychotherapy, re-connection with nature and greening the city, these projects play an important role in the city as well as in the creation of new identities and a sense of belonging for urban dwellers. However, despite their success and their growing popularity, municipal allotment gardens are considered a short-term action of social

T. Anthopoulou (✉) · S. Nikolaidou
Panteion University of Social and Political Sciences, Athens, Greece
e-mail: antho@panteion.gr; so_nikk@hotmail.com

M. Partalidou
Aristotle University of Thessaloniki, Thessaloniki, Greece
e-mail: parmar@agro.auth.gr

M. Petrou
National Centre of for Social Research (EKKE), Athens, Greece
e-mail: mpetrou@ekke.gr

C.-T. Soulard et al. (eds.), *Toward Sustainable Relations Between Agriculture and the City*, Urban Agriculture, https://doi.org/10.1007/978-3-319-71037-2_11

policy rather than a long-term sustainable urban planning strategy challenging the conventional modes of land management and governance in Greek cities.

Keywords Motivations · Governance · Sustainability · Crisis context

11.1 Introduction

Urban agriculture (UA) has rapidly expanded during the last decades in the industrialized West, in different associative forms including allotment garden associations, community and municipal gardens and other social farming projects. Nowadays it has been widely legitimated in public discourse for its multifunctional role in urban sustainability (Aubry et al. 2012; McClintock 2010). A growing number of cities recognize the socio-economic and ecological benefits of UA and are becoming more likely liable to include agricultural activities in urban planning.

Nevertheless, the recent economic recession in the European South gave a new impetus to the development of urban gardens highlighting the role that UA may have in addressing urban food insecurity and socio-economic deprivation in an increasing part of society (Bell et al. 2016). These contemporary alarming phenomena recall the historic development of UA initiatives in times of crisis (World Wars, economic crises) both in North America and Europe (Mok et al. 2013).

This is the case of Greece where UA has never been a tradition. Urban allotment gardens are a novelty to city life closely associated to recent economic turmoil, which generated food indigence in many urban households. The lack of previous experience in urban gardening is mainly explained by the fact that urbanites have never broken their ties with the countryside, given that rural exodus and urban concentration processes are relatively recent in the country; i.e. after World War II (Partalidou and Anthopoulou 2015). In this context, a large part of urban dwellers have smaller or greater access to farmhouse products and locally processed food (olive oil, cheese, pasta, fresh vegetables etc.) through kin or friends networks keeping in touch (real and symbolic) with rural settings as well as wholesome and tasty food and authentic savors "of the village".

The very first individual urban gardens have emerged as grassroots initiatives in Athens in the early 2000s and are limited to small-scale city lots. Amidst economic expectations and euphoria during the preparation of the Olympic Games 2004, quite a few citizens began to challenge the social utility of costly mega-projects within the city and on its outskirts. Rising concerns about the urban model and the future of public space were fueled by the intensification of urban sprawl phenomena at the expense of peri-urban agricultural land and natural space in the Athens periphery. This is the example par excellence of the expropriation of large areas of the traditional vineyards and olive groves in the Messogheia plain (Attiki Region) in order to establish major large-scale infrastructure (new international airport-2001, Olympic Games infrastructures-2004, peripheral roads and major highways, etc.), which led to a great loss of farmland and natural landscape (Nikolaidou 2011). At the same time, further construction of the remaining vacant land in the already dense inner city (sports facilities, shopping centers, residential areas), as for example the case of the Municipality of Maroussi, also illustrates the trends discussed.

Furthermore some other individual and still small-scale grassroots initiatives of community gardens appeared during 2008–2009, reflecting the emerging societal concerns about urban land uses and planning. In the form of "green guerilla" these initiatives put forward an agro-ecological perspective as an alternative to the dominant planning model and real estate market-driven practices in Athens and in Thessaloniki (in northern Greece).

The outbreak of the financial crisis shortly thereafter (2009) and the activation of "indignant citizens" against the privatization (through fast-track processes) of public land and natural resources further boosted these gardens of the green guerilla type. During a time of deepening crisis, various alternative food networks and civic grassroots movements appear to supply fresh affordable food to urban dwellers suffering from (neo)poverty: the first CSA baskets have been appearing in the metropolitan Athens area; the "without intermediaries" movements have been spreading all over the country after the so-called popular activist initiative of the "potato movement" (2012) cutting out middlemen in the food chain through direct sales; NGO's, citizen's collectivities and municipal initiatives take actions to relieve vulnerable social groups through soup-kitchens, social groceries etc. In the general context of a multidimensional crisis (financial, political, economic and humanitarian) followed by hard austerity measures, the focus of urban agriculture tends to shift from claiming free spaces to claiming subsistence farming for livelihood. By taking into consideration that almost one quarter of the Greek population lives under the threshold of poverty,[1] the food and nutritional questions become urgent ones. This trend towards social and food security functions does not come into conflict with previous aspects of urban agriculture but rather creates complementary and synergistic effects (Anthopoulou et al. 2013).

This chapter explores the emergence of urban agriculture initiatives in Greece by first providing an overview of the phenomenon in the light of the economic crisis. It then focuses on municipal allotment gardens, as the most widespread and popular form of UA in the country. The main question addresses the sustainability and the durability of the municipal projects particularly regarding governance perspectives of new urban gardening practices at this time of crisis.

[1] In 2013, 23.1% of the total population was at risk of poverty; the unemployment rate was 27.2% and people living in households with very low work intensity amounted to 19.6% of the population aged 18–59 years old, according to "Statistics on Income and Living Conditions 2013", Hellenic Statistical Authority, Press Release, 13 October 2014. In 2013, Greece holds the third highest rate of poverty and social exclusion in the EU-28 after Bulgaria and Romania (Ziomas 2014).

11.2 The Diversity of Urban Agriculture Practices in Times of Economic Crisis

A cursory glance at the timeline of the emergence of urban agriculture in Greece (Fig. 11.1), reveals its diverse types, including community gardens under the "guerilla gardening" paradigm (Adams and Hardman 2013) and institutional allotment gardens. They all meet different needs, ideologies and political positions of the involved actors.

Elaborating on the timeline we can find the most emblematic case of guerilla gardening practices, the self-managed "Farm of Agros in Ellinikon"[2] in the metropolitan area of Athens. The farm covers 2.6 ha including a small vegetable garden and a newly planted olive grove of 1400 trees. It was created in 2011 by a community group of political and ecological activists as part of the struggle to prevent the large-scale privatization and commercialization of the former international airport of Ellinikon (including the former USA military base). Another popular example of a grassroots community garden in northern Greece is the PERKA (*PEriastikes KAliergies* in Greek meaning periurban agriculture) on the fringe of the city of Thessaloniki. Created in 2011,[3] it is located in an abandoned military encampment (68.9 ha).

During the same period (2011–2012) we find the first institutional examples of urban gardens deriving from environmental NGOs (e.g. the environmental awareness Park *Antonis Tritsis* in Greater Athens[4]) and other institutional entities such as the School of Agriculture in the Aristotle University of Thessaloniki[5] (AUTh initia-

[2] The whole domain is considered as the last unbuilt big public land in Greater Athens (620 ha) in the southeast coastline of Attica. Initial government plans during 2000s was the creation of a high green Metropolitan Park but debt crisis in 2009 foiled this perspective. The whole domain has been internationally promoted through fast track procedures as one of the largest planned real estate development project in Europe. The basic aim of the collectivity of about 100 growers and volunteers is not food production per se but an educational and demonstrative social farming project; namely based on organic farming within agro-ecology principles, maintenance of seed bank and free exchange of traditional seeds to interested individual growers and urban collectivities and support to school gardens and other community gardens. http://agroselliniko.blogspot.gr/

[3] PERKA was initiated by urbanites who share community volunteer work and benefits of "growing your own food organically" inspired by the principles of collectiveness and biodynamic farming. The first group began cultivating a small part for members' needs and supporting vulnerable social groups. As more interested people are joining the initiative, nowadays there are five PERKA groups of 30–40 gardeners each in individual and common plots and about 200 people in total. http://perka.org/

[4] http://parkotritsi.gr/

[5] The allotment gardens in AUTh are considered as a very successful initiative covering a huge demand of urban dwellers in the big center of Thessaloniki. It is worth mentioning that during the first open public call for a plot the registrations came up to 4.700 applications. Gardeners pay a minimum annual fee and everyone is eligible to apply without any social criteria. The principles of organic production are followed and users are obliged to have a series of seminars in regards to sustainable practices. Students and staff from the School of Agriculture provide everyday agronomical advice and support the gardeners with the help also of a bio-agronomist. People are

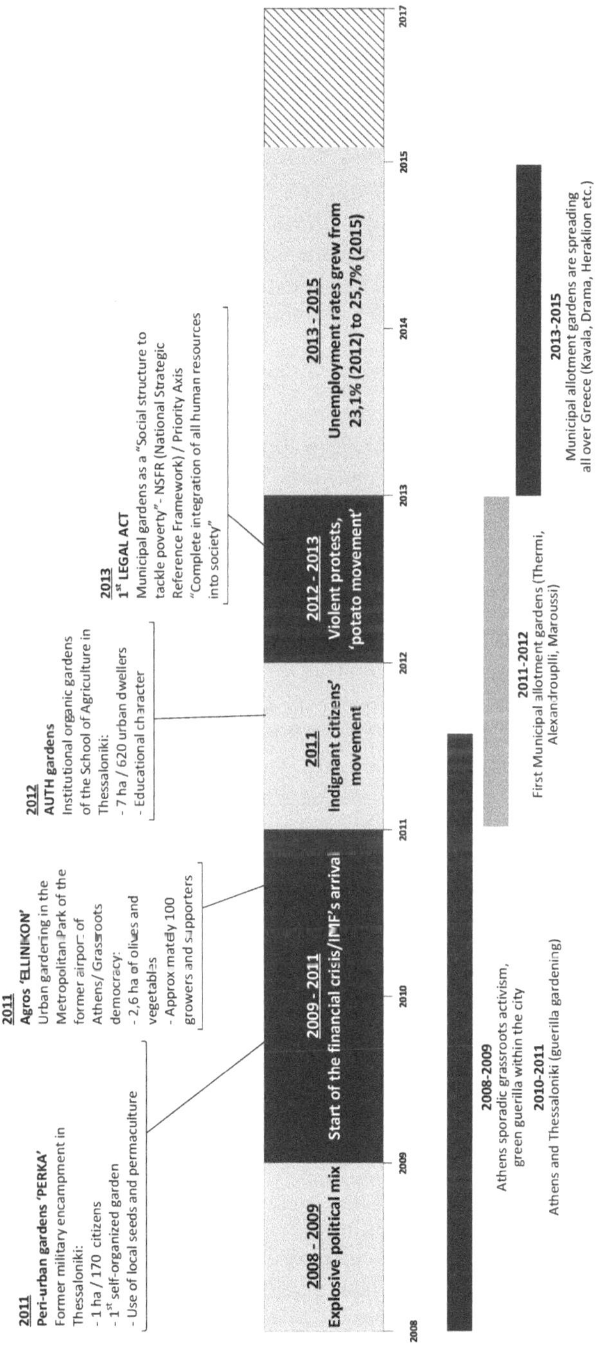

Fig. 11.1 Timeline of the eemergence of urban agriculture initiatives in Greece

tive). They aim at a sustainable city responding to the societal quest for a healthy diet and access to fresh local food while promoting educational benefits for all social groups, especially young generations.

The great boom of UA in Greece was noticed after 2012 with the emergence of municipal allotment gardens. They represent the most standardized and popular type of urban agriculture in Greece that was established as a top-down initiative by local authorities. The first gardens emerged during 2011–2012 in several cities all over the country (e.g. Alexandroupolis, Thermi, Larissa, the Greater Athens area etc.) as a spontaneous initiative of municipalities within social policy schemes to tackle poverty and depression of people affected by the economic crisis and austerity measures after the implementation of the fiscal adjustment program. Despite the fact that two of the pioneers (Thermi and Alexandroupoli) have initiated the idea as an urban greening policy tool, municipal allotment gardens primarily address vulnerable social groups with a view to easing the household food expenses and creating a social safety net in neighborhoods.

Therefore, since 2013, municipal gardens have been proposed by the Ministry of Labor and Social Solidarity to operate within the National Strategic Reference Framework (NSRF 2007–2013, co-financed by the European Cohesion Fund) under a specific action "Social structures for immediate fighting against poverty".[6] In other words, they are clearly conceived and organized by local authorities as a social policy action rather than an integrated long-term scheme for sustainable urban development (e.g. landscaping, waste recycling, reduction of carbon footprint and urban heat island effects, urban food planning). Nevertheless, the inclusion in the NSRF constitutes the only explicit reference to urban agriculture in public policies and therefore its sole legitimizing element in the absence of a specific urban planning regulation.

Municipal allotment gardens are considered as a social innovation in Greece at this time of economic and social insecurity. They are well embraced by citizens as introducing new ways of governance and dealing with city problems and as alternative spaces in city neighborhoods and a way out of the multiple effects of the recent crisis. The growing number of municipal allotment gardens highlights their social and symbolic value in light of the economic and humanitarian crisis.[7] For the first time the food security issue has been brought into the urban policy agenda in Greece, a country with strong ties –as mentioned- with rural areas.

---

responsible of bringing their own tools but are provided with free water by the University, which must be conserved. Seeds can also be provided, by the School, at a very low price. Today, more than 620 plots (of 100 m^2) provide almost 1000 urban dwellers with the joy and fulfillment of "getting your hands dirty" as well as a great amount of fresh vegetables for their family and friends. http://eco.auth.gr/wordpress/?page_id=3425

[6] More precisely, municipal allotment gardens are one of the eligible categories of public supported social structures -in addition to social groceries, homeless care, social dispensaries and pharmacies, and time exchange banks- to combat crisis stress and social exclusion aiming at promoting all citizens' integration into a society of equal opportunities.

[7] See indicatively the New York Times article: Alderman (2013) "More children in Greece are going hungry" on http://www.nytimes.com/2013/04/18/world/europe/more-children-in-greece-start-to-go-hungry.html?ref=global-home&_r=2&

In this political and economic context, the key research question relates to whether urban allotment gardens in Greece as a circumstantial effect of the economic crisis are sustainable over time with regards to the willingness of local authorities to legitimate UA and decide on an overall sustainable urban plan embedded with the values of the multifunctional role of UA. Nevertheless the question also remains open for urban dwellers in terms of their vision and willingness to support the gardens beyond the discourses on food security. The aforementioned questions are addressed through the field research in three municipal allotment gardens in northern Greece and in the metropolitan area of Athens. The research problematic focuses on three specific topics: (i) the local authorities' aspirations and the institutional and political context of the establishment of allotment gardens, (ii) the motivations of those who apply for a municipal plot and the lived experiences of gardeners after the first harvest, (iii) the sustainability and future prospects of these "crisis gardens" within national public policies.

11.3 Exploring Municipal Allotment Gardens in Greece. Outcomes from Field Research

11.3.1 Material and Methods Based on Three Case Studies

Municipal allotment gardens spread all over Greece with more or less the same profile and basic characteristics. According to empirical data from field research, they are set up in small areas in the urban area (with plots of 25–50 m^2) or larger ones in the outskirts (with plots of 50–70 m^2). Users have free access to land resources and water and some basic infrastructure (sprinkler system, fencing, soil amendment) provided by the Municipality. In some cases there is technical support by an agronomist and it is oriented towards organic vegetable production for self-consumption (not for sale). Municipalities have used social criteria for selecting the gardeners (pensioners, unemployed, low income, single parent families, etc.). Additionally, they try to connect the gardens to other social initiatives by asking, on a voluntary basis, from the users to offer a percentage (10–15%) of their production to the municipal social grocery, or even by giving away a plot to volunteers that grow vegetables for those that cannot.

At the helm of this innovative action appear to be the Municipalities of Thermi and of Alexandroupolis, the first established allotment gardens in the country. Both cases are introduced as a "top-down" initiative and are located in the urban-rural fringe of both cities (73,000 inhabitants in Alexandroupolis, 13,000 inhabitants in Thermi, Population Census 2011). With that in mind, we selected these two case studies in order to explore aspirations, objectives and the political context of their establishment by the municipal authorities, on the one hand, and on the other motivations and lived experiences as expressed by the urban growers. The third case is also a pioneering municipal garden, the difference being that it is located in the inner city of the Greater Athens metropolitan area, that of the Municipality of Maroussi (72,300 inhabitants). Another differentiation in this case study is the top-down decision of the municipal council to set up the vegetable garden, reconciled with an existing bottom-up demand of citizens to preserve vacant lots in the dense urban fabric for social uses and stabilize the neighborhood (Fig. 11.2).

As already mentioned, the key research question addresses the sustainability and the duration of the allotment garden projects on the part of both the municipal authorities (i.e. integration of urban agriculture in the urban plan) and the growers (i.e. commitment in farming and agricultural good practices).

In doing so we used several methodological tools: Key informants were used in all three gardens in order to provide insights on the history of the establishment of the garden (they actually were the mayors, the municipal agronomists, community

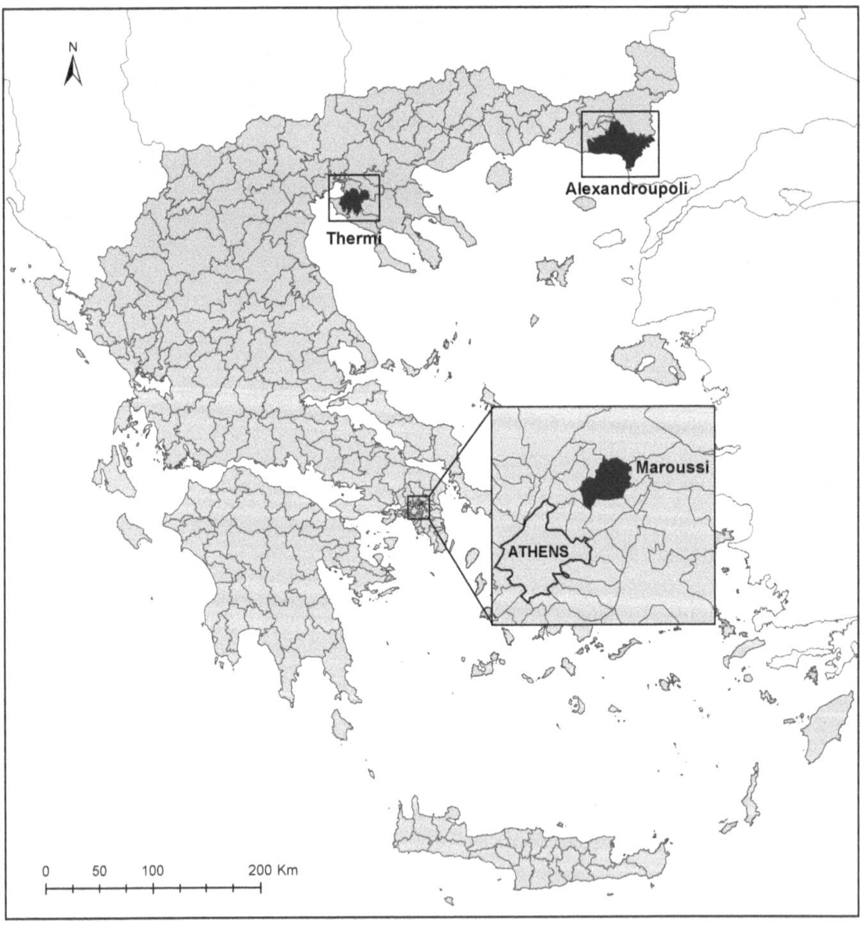

Fig. 11.2 Location of the municipal gardens of the study

associations' representatives etc.). Subsequently, a survey with a fixed questionnaire was used with urban dwellers that had applied for a plot at the two gardens of Thermi and Alexandroupolis. This part of the research took place before harvesting. Nevertheless, in order to elaborate on the lived experiences (post-harvest) we used focus group discussions with growers in all three gardens. At this point we must note that the survey on the growers (pre-harvest) was not conducted in the case of Maroussi because during the time of the research they had already been appointed to their plots and started growing.

Interviewees for the survey came up to 63 in Alexandroupoli and 141 in Thermi (which was the total number of all the people that had applied-by that time- for a plot). This cohort of urban dwellers was identified by the local agronomists involved in the projects through a list of names of people that applied for a plot responding to the open call of the municipality. The growers for the focus group discussions were also identified by the local agronomists based on a list of criteria that we provided them (age, sex, rural background, educational background etc.) in order to have a differentiated profile.

Our data were subjected to both quantitative and qualitative analysis. The data that derived from the questionnaires were subjected to a statistical analysis using frequency tables and descriptive statistics to elaborate on the motives. With regard to the qualitative data, discourse analysis was used to answer our main research questions and to conceptualize the findings.

11.3.2 The Municipal Allotment Gardens in Alexandroupolis and Thermi in the Rural-Urban Fringe: An Initially Greening Policy

It was rather evident from the point of view of the public bodies that the city needs a greening policy for both environmental and educational purposes. According to the results of our field research both Municipalities had several environmental projects going on before the idea of the garden. In the case of Thermi, a home-composting project was being undertaken with bins distributed free of charge to households in order to stimulate sustainable attitudes of city dwellers. From this compost project emerged the idea of a small garden, which could be the recipient of all the produced humus. In the case of Alexandroupolis, the vision of a "bioclimatic shielding" of the city was the stimulus for the establishment of the garden. They envisaged the garden as a continuum of the city; more or less a dreamed utopia.

Later and as the economic crisis started to influence citizens both projects were embedded with other aims such as the social policy aim of food provisioning for vulnerable social groups or the socializing aim and reinforcing self-esteem to battle against psychological stress and other collateral effects of the crisis. In fact, the economic crisis played an important role in the decision of the municipalities to re-enforce the vegetable gardens given the worrying phenomena of neo-poverty and malnutrition of a growing part of their population. Another motivation expressed by local authorities, according to our key informants, was also to "cultivate social responsibility and solidarity". The latter was achieved by requiring from growers to offer 10% of the crop yield to the municipal social grocery. Last but not least, the environmental concerns, which were from the start a priority, were enhanced in the practices of the gardener towards organic growing.

With regard to the motivations of the citizens, the results of our statistical analysis reveal almost the same priorities. In both cases of Thermi and Alexandroupolis, urban dwellers were motivated primarily by the expectation of ensuring fresh and quality-organic- food for the family. In fact, a percentage of 72.34% of the applicants for a plot in the municipal allotment garden of Thermi (Fig. 11.3) were driven by this motivation (either as a first priority motive or second or even third). The same also holds true for an even higher percent of 88.89% in the case of Alexandroupolis (Fig. 11.4).

The motive of economic hardship and the need to minimize expenses for the daily meals of the family was expressed by an overall percentage of 44.68% for Thermi and 68.25% for Alexandroupolis. Other motivations (such as psychotherapy, community building, re-connection with nature and greening the city) were less prioritized by the applicants for a plot in both gardens.

At this point it is also crucial to provide the profile of the people involved in our research (Table 11.1). According to the descriptive analysis, the majority of the respondents were male for both cases; in their mid-40s for Thermi whereas for Alexandroupolis they come from mixed age groups. But this was indicative only of

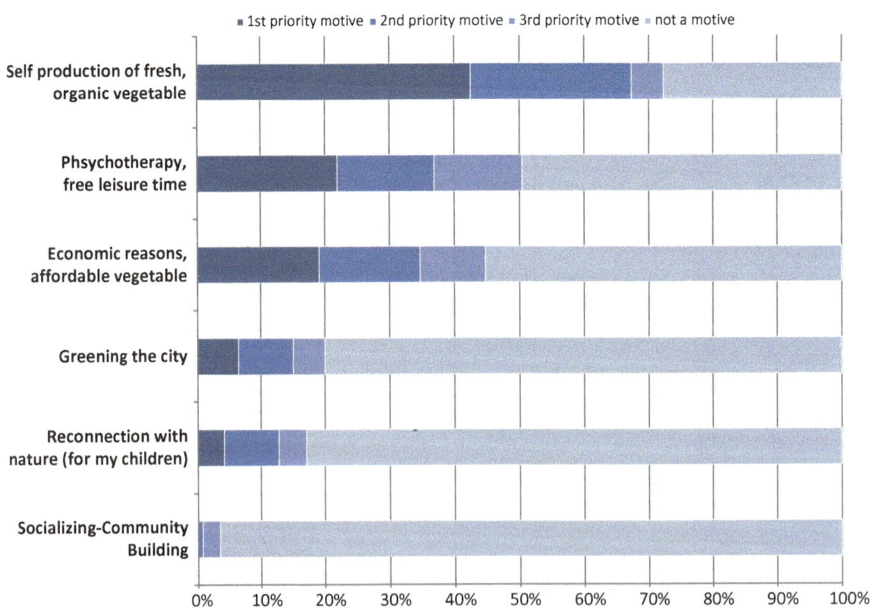

Fig. 11.3 Urban dweller's motives in the case of Thermi (pre-harvest period)

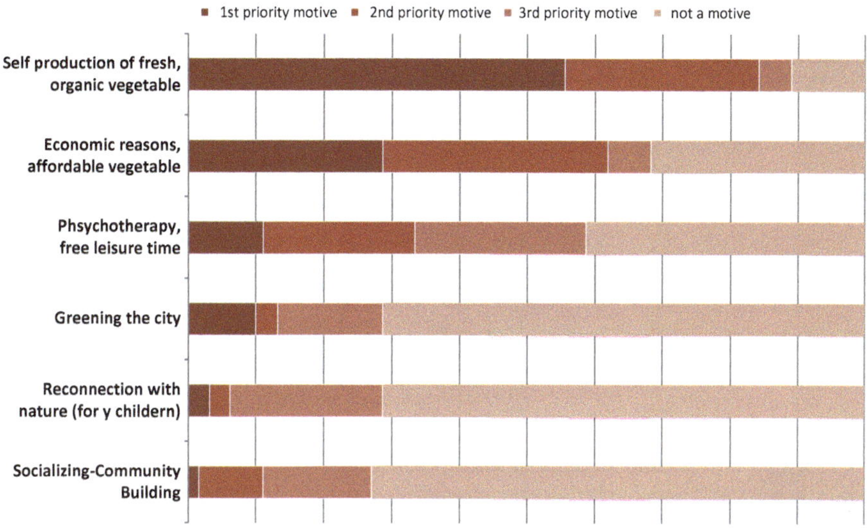

Fig. 11.4 Urban dweller's motives in the case of Alexandroupolis (pre-harvest period)

Table 11.1 Profile of the people involved in the research

Variable	Thermi	Alexandroupolis
Sex and age	Male 62.4%, mid-40s	58.7% male, mixed age group
Family structure	Single(s) 7.1%	Single 14.3%
	66% had a family (16.2% had more than four members)	43% had a family with minors
Educational level	32.6% high school	36.5% high school
	35.5% university degree	22.2% university degree
Employment	32.6% in public sector	15.9% in public sector
	19.1% in private sector	17.5% in private sector
	5.7 businessmen/women	9.5% businessmen/women
	24% unemployed	22.2% unemployed
	13.5% pensioners	27,0% pensioners
	2.1% housewife	7.9% housewife
	2.8% student	
Experience of gardening	Yes 61% (mainly from practicing gardening in their village)	Yes 68.3% (mainly from practicing gardening on balconies and in kitchen gardens in the city)

Source: Anthopoulou et al. (2013)

those who applied for the plot on behalf of the whole family (the male applied for both spouses). Single people were also found in both gardens. With regard to the educational level it is worth mentioning that in Thermi we found a slightly more educated group of people, more public servants and fewer pensioners. Finally, both research groups had prior experience of gardening, either from their home balconies and kitchen gardens in the city or from agricultural activities in their village of origin.

Then, using the aforementioned different profiles we tried to find people with different characteristics both male and female, from all age groups, pensioners, unemployed *etc.* in order to establish our focus group discussions. The outcome of this part of the research highlighted post-harvest satisfaction (as they were reached after their first harvest) more than the major motivations before the actual experience. What people verbalized during the discussions were the lived experiences of their everyday life and encounters in the garden.

Feelings that were unknown before with regard to where and how food is grown, a sense of self achievement and getting close to nature were articulated by many gardeners regardless of their different characteristics. What growers mostly emphasized were the pleasure of growing your own food and learning from nature. "*It's an unusual feeling for me and for the whole family. You can't imagine how important it is for a child to realize for the first time that there is another way to get vegetables instead of going to the supermarket and picking them based on their nice packaging and their positioning on the shelf or just choosing the shiniest from the shelf, weighing them, paying for them and taking them home. In the garden you can see the cucumber that you have nourished from seedling, you can cut it and eat it regardless of its size and shape, you don't even need to rinse it off with water. Even for me the experience is rather new, despite the fact that I come from a village, because it was my parents that were responsible for everything. I already knew that feeling when you pick peppers or tomatoes with your own hands, but now the difference is that I have the total responsibility for growing the plants from scratch. This is the most important benefit for me personally*", mentioned a young father from Thermi.

Urban dwellers started cultivating based on 'learning by doing' which was in the very beginning consuming and frustrating but actually gave them great satisfaction and self-esteem. What a young woman from Alexandroupolis, unemployed at the time, mentioned is indicative: [...] "*I had no idea how to grow vegetables. I did everything on my own...I would have appreciated it at the beginning if someone-an agronomist for example- just gave me some tips about how to start. I remember when I was putting the seedling in the soil... I was wondering what the outcome will be. God knows! When I saw the produce for the first time I said thank God I made it! It is a great satisfaction and self-achievement. Especially nowadays as we are considered as worthless and incapable of finding a job and offering something to society*".

In this process of "learning-by-doing", compensation is not related to the production cost of vegetables but with feelings of self-esteem. [...] «*if you asked me what was the most amazing thing, I would tell you that it was the first time I cut a melon. It is the best experience ever. It makes a sound as you cut it off and it needs a special technique too. But if you ask me about the economic benefits I will most certainly tell you that we spend more money than we will ever save. But the fact that you eat something that you grew with your own hands, it's worth it and you can't put a price on this value*» (male gardener in his mid-40s, single, Alexandroupolis).

We have to note however, that despite all the merits, growers had to confront several problems and difficulties. Some of which were the lack of competence in growing techniques and knowledge of basic agronomical practices especially with regard to organic production. The agronomists from the municipalities could not always support them given that they were mostly oriented towards mainstream practices. All the above made the growers turn to other knowledge networks such as the Internet, social media (blogs for exchanging problems and solutions) and also use empirical knowledge and practices of their parents in farming and overcoming rural hardships.

Despite the technical difficulties, it seems that enthusiasm is prevailing over the inconvenience of the amateur grower. The most surprising outcome of the lived experiences in the garden was that the garden was perceived as a place of growing not only vegetables but «growing» also friendships, a sense of community and a sense of safety amidst a crisis of values and a crisis of identity (Delshammar et al. 2016). "*I never imagined that I would meet so many people here. There is nothing more satisfying than that. The other day I was looking for some tools and everyone was more than happy to let me use their own. That is the point of a community garden: to help each other, provide for each other, not only physically and materially and just exchange gardening knowledge but create common values and solidarity*", said a middle-aged housewife from Thermi. A newly unemployed responded from Thermi added: *«It's been some time since I lost my job and for me the garden is a place where I feel safe. I come here, to work in the garden; sometimes I just sit and think, or make small talk with other gardeners. It's a way-out for me, sometimes we even drink ouzo and have a nice time here without any concerns of what will happen in life»*. Finally, the garden represents a vivid meeting place, a way-out from crisis stress (Partalidou and Anthopoulou 2015). Benefits from growing your own food are clearly higher than just covering nutritional needs. It seems that in times of crisis escaping from everyday stress is a more valuable challenge than food production *per se*. *«I love it here. It's a place where you forget all other matters-problems. Sometimes I am so tired that I don't want to get out of the house. But once I cross the gate I forget my tiredness and instead of staying one hour I usually get off track of my household routines and stay till the sun goes down"*, mentioned another housewife from Alexandroupoli. *«I too am more than happy. I was thinking of coming once or twice a week but now I come two times a day, once in the morning and then late in the afternoon. Honestly, I spent the whole summer like that. My friends in the 'kafeneion' (coffee-shop) were looking for me…[laughs]"*, added a friend of theirs, a pensioner.

11.3.3 The Municipal Garden of Maroussi in the City Block: The Mobilization of Local Residents to Stabilize the Neighborhood

Our third case study is the one in Maroussi, a relatively green suburb in the north of Athens. It is a municipal allotment garden slightly different regarding stakes and motivations of all involved actors – including the neighborhood residents, the municipal authority and the beneficiary growers. It is located within the urban fabric representing an innovative case where neighborhood's activism was reconciled with the municipality's social action. The garden was established in 2012 on an abandoned municipal land (1500 m^2), which over the years was downgraded and turned into a dump. This unbuild plot which was originally intended for green infrastructure and leisure was a subject of controversy for several years –in the early 2000s- mobilizing local residents in order to prevent the construction of a multipurpose municipal centre.

This municipal garden is gaining particular symbolic value given that the city of Maroussi has been marked by intense urbanization processes since the 1990s. This transformation is due to middle class suburban housing development and also to the establishment of the headquarters of large companies, banks and big shopping centers. The municipality also hosted some mega Olympic projects of 'Athens 2004' (e.g. the Olympic stadium and residential complexes) that led to a further building boom, gentrification processes and the rise of real estate market prices. Amidst the financial and economic crisis, the Mayor proposed to create a vegetable garden on this site. The project has paralleled the collapse of the real estate market, which thereby suspended any further construction activity in the country. At the same time, it was in line with the dominant public discourse of coping with neo-poverty phenomena in cities. As stated by the Mayor at the opening of the garden: "*At a time when the economic crisis leads many of our fellow citizens to despair, the Municipality of Maroussi offers opportunities for activation and a creative way out, while providing a safety net for socialization and social care for the residents of our city.*"

The vegetable garden of Maroussi is otherwise typical of municipal garden allotments in Greece. It contains 35 allotments of 25–30 m^2 and during our field research 31 growers were already on-site growing vegetables. Out of them 17 are male and 14 female. However, if we consider that only 5 are single it can be in fact characterized as a family garden, similarly to both Thermi and Alexandroupolis. Among the growers 13 are pensioners, 12 workers, 4 housewives and one (1) unemployed. Access to land is free, basic infrastructure is provided by the municipality, farming practices are organic, and the beneficiaries are mainly selected using social criteria with priority given to people living near the garden.

According to our focus group discussion with the growers in Maroussi it is evident that their satisfaction stems from self-production of fresh and quality food, physical exercise, conviviality, and escape from the burden of the crisis. As a male pensioner former policeman, stated: "*I come here every day with my wife. It's a pleasure, a physical exercise, a way out; despite the everyday problems we face regarding our health or family matters, as we get older. And it also reminds us of our childhood...when I was a child, in my village; I used to help my parents in the fields, as all children used to do back then […].*

His wife, a housewife, adds: "*I also come from a village but my parents were not involved in agriculture. Personally, it was a dream come true to have a piece of land of my own, even that small plot of 2m-x-2m. For a while we had lost touch with earth here in Athens...but now I couldn't be happier! I grow my own vegetables and take care of them as if they were my children. As they say the "the best fertilizer for the plants is the farmer's shadow*".

Economic alleviation was not articulated as a major a priority as the food quality. *"Economically, no, it's not a relief; if we calculate the manure, seedlings [...], it all adds up. But I eat fresh pure food without pesticides; I give it to my children and my grandchildren, sometimes even to my neighbors"*, a female pensioner former school-teacher, mentioned.

Despite their contestation over the economic gains from the garden, everyone agrees that they have many indirect economic benefits given that self-production made them better appreciate the value of food while reviewing their attitudes as consumers by purchasing only the necessary, seasonal and local, trying at the same time to avoid food waste.

For a pensioner, former bus driver-a 'fanatic' grower in the garden- UA provides economic benefits as he saves by not taking coffee at the coffee shop and from stress relief medicines! As he explained, he would have to spend a fortune on heart medication under the pressure of the crisis and uncertainty. Instead, he comes to the garden everyday and also helps other growers!

Even if economic relief is not a priority motivation for many gardeners of Maroussi, economic support of vulnerable citizens comes through social solidarity and activation of the community bond: "*My wife knows people in need from our parish; sometimes we give them some of our produce; if you live in a household of two people you can't consume that much*" (grower, civil servant).

Despite all the merits of the garden, growers face many problems and conflicts. The garden is undermined by the water shortage during summer time, which is also an indirect effect of the crisis. As our key informants explained, the municipality vehicle that brings water to the garden broke down in early summer of 2014 and the municipality has no available funds for the purchase of a new one. In addition, due to the layoffs of personnel –including drivers of the municipal car fleet- (following the Economic Adjustment Program) the water supply was not regular (both in time and frequency) so growers had to be rather frugal and commonly decided on the use of spare water for the plants. Today, they often refer to "the water war" that created conflicts and disrupted friendly relationships they had until now. "Whatever you say, I believe that the garden is a reflection of our society. With people that are good, earnest but also wicked and vicious…you see, I work all day and I come to the garden late at night. I can't be here during the water supply from the municipality. So you can imagine that by the time I come to the garden all the water has been used. People do not treat it as a common good. I spend all my time here fighting with other people about their bad behavior I am so frustrated", stated a young male gardener, working in the private sector.

Last but not least, we must also discuss the matter of the temporal character of the garden which impedes its institutional consolidation. As another gardener said (pensioner, former electrician): *«I come every day twice, even three times, a day. The garden is a vital need, I have invested much here. I brought soil, I put fences,* I have even made some customizations to store water so that I wouldn't have to depend on the municipality. I can't imagine that one day they will come [from the municipality] and force me to leave this garden in case they intend to change land use. So much labor, so many expenses". It is evident that the lack of an institutional framework is very crucial and as already mentioned holds true for almost every municipal allotment garden.

11.4 Discussion: Governance Issues and the Question of Durability of Municipal Garden Allotments in Crisis Context

Urban agriculture in Greece in the form of municipal vegetable gardens is a relatively recent and ongoing phenomenon; hence it is not possible to draw safe conclusions about its dynamics and future perspectives. Evidence so far shows that municipality-driven initiatives are considered as 'mainstream' and short-term projects that emerged in a context of economic crisis and consequent real estate collapse. A number of key constraining factors in terms of urban policy and governance practices have strong sustainability-related implications.

It is worth underlining that the common patterns of this type of gardens were initially developed by local authorities through spontaneous and rather informal processes without any explicit local or national strategy. However, later on they started gradually to gain some kind of political recognition via public support from municipal and national budgets that refined initial public interests and motivations. Fieldwork research showed that although few municipalities were primarily inspired by environmental and educational objectives (organic production, composting of organic urban waste, green spaces etc.), yet, very soon the majority of initiatives were integrated into a socially-centered policy scheme. Therefore, given the worrying symptoms of neo-poverty amidst the economic hardship and severe austerity measures, gardens were granted significant resources from the European Cohesion Fund (National Strategic Reference Framework-NSRF). Overall, this fund is expected to support moderating the impacts of the fiscal adaptation program, which dictated direct salary cuts, high taxation, the shrinking of social security benefits, redundancies, high rates of unemployment and unprecedented economic and social deprivation especially in cities. Inevitably, priority was given to vulnerable social groups by municipalities in order to alleviate families' budgets on food expenditure and create a social safety net in city neighborhoods.

Nevertheless, though municipal gardens were predominately developed to face the economic hardship they have eventually evolved from "crisis gardens" to spaces that mainly provide private food safety, social conviviality and psychological relaxation to growers. Notwithstanding the impoverishment of urban households, urbanites expressed the need to grow their own vegetables so as to have fresh and healthy rather than cheap industrialized food, which one may find in supermarkets anyway. Despite the fact that the production cost in the garden is quite significant-compared to vegetables from the market- from a socioeconomic cost-benefit analysis point of view, the overall benefits (social, emotional, psychological, etc.) for themselves, their family and the local community are higher than the costs and the commercial value *per se* of their vegetable production, thus encouraging them to continue growing.

Despite the multifaceted benefits experienced by urban growers, long-term viability and sustainability of municipal gardens is rather inhibited by a lack of institutional consolidation and legitimization of planning and policy tools. Based on our fieldwork there is no official integration of these initiatives in the urban planning processes in terms of land-use or zoning. This means that municipal gardens are promoted via informal land management mechanisms, which do not include them in the planning documents as a classified land-use category, thus they are not considered as a permanent urban function. On the contrary, increased temporality and informality of municipal practices is significantly enhanced by a period of slow real estate development and financial difficulties of the municipalities (Fritsche et al. 2011; Fuhrich and Goderbauer 2011). Under such circumstances, small-scale urban gardening preferably established on city-owned vacant land is a way for the municipalities to re-vitalize neighborhoods in inner-city or peri-urban areas. These interventions, as in many other cases all over Europe, demand minimal public support while at the same time they re-activate and maintain unused and un-built land resources awaiting construction. However, their ephemeral character facilitates the future use of the property for more competitive uses; thus they might disappear from one year to another.

That said, despite the success and the growing popularity of municipal vegetable gardens in Greece, the main sustainability challenges relate to the short-term character of the actions, the informal land arrangements as well as to the strong limitations of policy, planning and governance at institutional level. According to Robineau (2015), such a mixture of informal and formal arrangements developed at the municipal level can often generate a mechanism that somehow regulates land access and creates a certain level of political and social visibility of municipal gardening projects. However, the lack of horizontal integration among sectoral planning practices related to agriculture (Cinà and Di Iacovo 2015) and other green infrastructure of the city constrain the future of these gardens. The current governance pattern tends to use these new growing spaces as a spontaneous and ephemeral municipal tool primarily for social policy while it ultimately depends on available financial and land resources. This signifies that urban gardens are not currently incorporated in a long-term sustainable urban planning strategy as integral parts of the city's sustainability fabric. At the same time, future prospects in terms of social sustainability are also challenged by the dynamics of other relevant social practices in a context of an increasing socio-economic crisis and lack of open spaces in urban areas. The growing pressure of citizen movements towards re-appropriation of open spaces and empty lots (Rosol 2010) creates new 'political spaces' in the cities (Cohen and Reynolds 2014). This transforms the conventional use and functions of public space as well as the conventional top–down approaches in decision-making and implementation of practices are questioned. However, the Greek case reveals a lack of a broader participation scheme of public and civic actors in urban agriculture policy-making and planning that restrict the development of democratic processes in terms of public space governance. As long as Municipalities do not engage a wide range of stakeholders, these programs could likely fail to reconcile with the demands of numerous existing bottom-up gardening initiatives or other social and alternative food networks. Such enlarged collaborations could be based on available resources and be complemented by an extensive network of community groups that could potentially foster the city's governance mechanisms in order to link food-health and poverty issues with urban sustainability.

To sum up, either driven by economic or real estate crises, these projects may play an important role in the city planning process due to their socioeconomic and environmental dimensions that influence urban landscape and quality of life. Re-appropriation of public spaces in the vicinity of housing areas could contribute to the improvement of the quality of life and enhance social interaction and cohesion among urbanites. However, in a context of increased temporality and adaptability in terms of land access, the questions of legitimization of planning policies, governance, negotiation of space and the right to the city as a public good remain crucial issues for the viability of the allotment gardens.

References

Adams D, Hardman M (2013) Observing guerrillas in the wild: reinterpreting practices of urban guerrilla gardening. Urban Stud 51(6):1103–1119. https://doi.org/10.1177/0042098013497410

Alderman L (2013) More children in Greece are going hungry. New York Times. http://www.nytimes.com/2013/04/18/world/europe/more-children-in-greece-start-to-go-hungry.html?ref=global-home&_r=2&. Accessed 27 Feb 2015

Anthopoulou T, Partalidou M, Nikolaidou S, Kolokouris O (2013) Urban agriculture: social inclusion and sustainable city. Case study of two municipal gardens in Northern Greece (Municipalities of Alexandroupolis and Thermi). Panteion University, Athens

Aubry C, Ramamonjisoa J, Dabat M-H, Rakotoarisoa J, Rakotondraibe J, Rabeharisoa L (2012) Urban agriculture and land use in cities: an approach with the multi-functionality and sustainability concepts in the case of Antananarivo (Madagascar). Land Use Policy 29:429–439

Bell S, Fox-Kamper R, Keshavarz N, Benson M, Caputo S, Noori S, Voight A (eds) (2016) Urban allotment gardens in Europe. Routledge, New York

Cinà G, Di Iacovo F (2015) Integrating top down policies and bottom up practices in urban and Periurban agriculture: an Italian dilemma, future of food. J Food Agric Soc 3(1):9–20

Cohen N, Reynolds K (2014) Urban agriculture policy making in New York's "new political spaces": strategizing for a participatory and representative system. J Plan Educ Res 34(2):221–234

Delshammar T, Partalidou M, Evans R (2016) Builiding trust and social skills in urban allotment gardens. In: Bell S, Fox-Kamper R, Keshavarz N, Benson M, Caputo S, Noori S, Voight A (eds) Urban allotment gardens in Europe. Routledge, New York, pp 342–363

Fritsche M, Klamt M, Rosol M, Schulz M (2011) Social dimensions of urban restructuring: urban gardening, residents' participation, gardening exhibitions. In: Endlicher W, Hostert P, Kowarik I, Kulke E, Lossau Marzlff J, Van der Merr E, Mieg H, Nutzmann G, Sculz M, Wessolek G (eds) Perspectives in urban ecology. Studies of ecosystems. Springer, London, pp 261–296

Fuhrich M, Goderbauer E (2011) Urban restructuring – making 'more' from 'less'. In: Muller B (ed) German annual of spatial research and policy 2010. Springer, Berlin/Heidelberg, pp 49–58. https://doi.org/10.1007/978-3-642-12785-4_5

McClintock N (2010) Why farm the city? Theorizing urban agriculture through a lens of metabolic rift. Camb J Reg Econ Soc 3(2):191–207

Mok H-F, Williamson V, Grove J, Burry K, Barker S, Hamilton A (2013) Strawberry fields forever? Urban agriculture in developed countries: a review. Agron Sustain Dev 34(1):21–43. https://doi.org/10.1007/s13593-013-0156-7

Nikolaïdou S (2011) Politiques de planification et logiques des acteurs locaux dans le contexte de périurbanisation du sud-est d'Athènes. Options Méditerr B 66:145–158. http://ressources.ciheam.org/om/pdf/b66/00801380.pdf. Accessed 24 Jan 2015

Partalidou M, Anthopoulou T (2015) Urban allotment gardens during precarious time: from motives to lived experiences. Sociol Rural 57(2):211–228. https://doi.org/10.1111/sour.12117.

Robineau O (2015) Toward a systemic analysis of city-agriculture interactions in West Africa: a geography of arrangements between actors. Land Use Policy 49:322–331

Rosol M (2010) Public participation in post-Fordist urban green space governance: the case of community gardens in Berlin. Int J Urban Reg Res 34(3):548–563

Ziomas D (dir) (2014) Country profile–Greece. Unpublished report for the European social policy network. National Centre for Social Research, Athens

Chapter 12
The Proliferation of Collective Gardens in Lisbon (Portugal) and Montpellier (France): Urban Residents Demand and Municipal Support

Pascale Scheromm and Guilhem Mousselin

Abstract Today there is renewed concern about agriculture's place in the construction of sustainable cities. The rapid development of urban collective gardens (also named allotments and community gardens according to the authors and the countries) provides an example that takes on an exceptional dimension in the context of a systemic crisis that is simultaneously social, economic, environmental and political. In this chapter, we illustrate the diverse forms and issues observed during our study of these gardens, and compare their resurgence in two European cities: Montpellier, France; and Lisbon, Portugal. This comparison allows us to address the specificity of these collective gardens, their dynamics and their governance methods. Their proliferation in these two cities reflects a strong demand by urban residents, and the developmental support of the municipalities.

Keywords Urban agriculture · Multifunctionality · Governance · Public policy

12.1 Introduction

For most of history, agriculture existed in cities, occupying spatial patterns that closely combined urban and agricultural space (Steel 2009). The industrial revolution created a separation between these spaces, reducing agricultural presence in the cities of developed countries. But a new emphasis on urban agriculture is evolving

---

P. Scheromm (✉)
Innovation, INRA, CIRAD, Montpellier Supagro, University of Montpellier, Montpellier, France
e-mail: pascale.scheromm@inra.fr

G. Mousselin
Bordeaux Montaigne University, UMR CNRS 5319 Passages, Pessac, France
e-mail: g.mousselin@gmail.com

C.-T. Soulard et al. (eds.), *Toward Sustainable Relations Between Agriculture and the City*, Urban Agriculture, https://doi.org/10.1007/978-3-319-71037-2_12

Box 12.1: About the Term Collective Gardens
The term « collective gardens » refers to all types of cultivated spaces where groups of gardeners operate (Pourias et al. 2015). Indeed these gardens are named differently according to the geographic context. So the term « collective gardens » is an umbrella term encompassing historical and more recent forms of gardens, like the French family and shared gardens, the allotment plots in the UK, the community gardens in USA and in Quebec, the hortas urbanas in Portugal.

through various forms according to countries, regions and cities. This interest is particularly evident in the concerns of citizens and elected officials involved in the proliferation of urban collective gardens (see Box 12.1).

The first collective gardens were established in the nineteenth century in the context of increasing industrialization, urbanization, and the accompanying rural exodus. They experienced another surge in cities during the two world wars, when the cultivation of urban spaces was encouraged by public officials to address the socio-economic issues of food production. But after 1945, the number of urban gardens collapsed (Consales 2004; Hallsworth and Wong 2015), and their function as a source of food became less relevant. Today, in an atmosphere of concern for food security, availability and accessibility, this function is once again a subject of interest (Duchemin 2013; Pourias et al. 2015). The collective gardens come in various shapes and forms, ranging from traditional family allotments, to shared plots dedicated to the social reintegration of residents in difficulty or seeking employment. They represent various functions (Duchemin et al. 2010), and potential components in urban planning. Since the 1990s, these gardens have been seen among the issues addressed in urban policies (Guyon 2005; Mousselin and Scheromm 2015). Beyond their function of basic production, they respond to concerns about the quality of life (Reyburn 2002), social links and integration, nature, and a reappropriation of the food supply (Scheromm 2015). Community gardens are often found in small interstitial urban spaces, and have captured the attention of scientists as well as public actors. Their reintroduction into urban space has become a common consideration at the municipal scale in Europe and North America (Mansfield and Mendes 2013; Ernwein 2014; Hardman and Larkham 2014; Vitiello and Wolf-Powers 2014; Huang and Drescher 2015).

The context of the current crisis (simultaneously social, economic, environmental and political) creates conditions favourable to the proliferation of urban collective gardens. This phenomenon has already been seen during difficult social and economic periods (Pudup 2008). The systemic crisis has repercussions in various spheres, including food supplies, the proximity of nature in urban life, and citizenship links to the political structure.

In order to illustrate the diverse forms, the issues and the dynamics of these urban collectivegardens, we compare two European cities, Montpellier in France and

Lisbon in Portugal. This crisis exists in both countries, but it is more present in Portugal, as can be seen in numerous social indicators: the rate of severe poverty in Portugal, for example, is almost twice that of France.

Our choice to study these cities was based on both practical and scientific criteria. We studied Montpellier and Lisbon within a French research program on sustainability of urban agriculture in Mediterranean countries (Durabilité des Agricultures Urbaines en Méditerranée, DAUME). The scientific interest of the comparison is to analyze the proliferation and the development's conditions of the urban gardens in two different geographical and social situations. What are the issues engaged in the establishment and functions of these collective gardens? What actors are involved?

We show that the development of the collective gardens operates under the pression of a citizenship dynamic supported by the municipalities.

As a first step, after outlining the nomenclature of the collective gardens in the two cities, we will study their physical importance and location in each city, and identify their specificities. Following this characterization, we will look at the processes and the actors found at their origins. In conclusion, these results will allow us to discuss the place these gardens occupy in municipal planning strategies.

12.2 Case Studies and Methodology

Montpellier, the second more important city of the Occitanie region, has a population of 275,318 which makes it the 8th largest city in France. Since the 1960s, rapid urbanization has exerted strong developmental pressure on open space. The city has a historical tradition as a university town, and now has an economy based on the tertiary sector, specialized in research, culture, tourism, and services. Today, Montpellier is proactive in developing environmental policies protecting biodiversity and preserving green space (Scheromm et al. 2014). Agricultural land only encompasses 7% or 435 hectares of the city's surface area.

Lisbon, the capital of Portugal, has 506,892 inhabitants. A cultural and cosmopolitan city, Lisbon also houses universities as well as numerous national and foreign institutions, and the most active European port on the Atlantic coast. Professional agriculture exists only marginally, but like Montpellier, the municipality is developing an environmental policy designed around the implementation of its green plan since 2007 (Telles 1997), and the revision of its municipal master plan in 2014.

Our fieldwork was conducted in two stages. First we identified, inventoried and characterized the collective gardens in each city. In order to determine the issues and objectives of the gardens, we then interviewed actors from institutions and associations, and others involved in garden management. The gardeners were questioned as well in an effort to determine their motivations. The results of these investigations on the importance and location of the gardens, their organization, the actors involved, and the governance processes employed were then compared point by point.

12.3 Collective Gardens: Spatially More Extensive in Lisbon Than in Montpellier

12.3.1 *Different Types of Gardens: Different Forms, Statutes and Nomenclatures*

In France, we examine two categories of urban collective gardens: family gardens and shared gardens (Fig. 6.1). Family gardens are created on "land divided into plots, that are then allocated to individuals for private gardening practices benefiting their personal needs or those of their family" (article L561-1 of the rural code). The sale of produce from the garden is not authorized. French family gardens are a carry-over from the industrial revolution, and have existed since the end of the nineteenth century. They have been studied more in the cities of Northern France (Cabedoce and Pierson 1996; Dubost 1997) than in Southern France, where they are more recent (Consales 2004). The gardens are characterized by their functions providing both food and leisure activities for urban citizens seeking to cultivate the land. They appear as an ensemble of plots, most often fenced, with a small shed, and connected to a public water system. The size of the plots rented to citizens for family gardening generally vary from 100 to 300 square meters. The shared gardens are a more recent form of collective garden. They are defined according to a law adopted in July 2007 as gardens conceived and cultivated collectively by the residents of a neighbourhood or village "with a goal of developing local social links through cultural or educational activities". Related to the North American community gardens, these shared gardens appeared in France in the 1970s (Pashchenko and Consales 2010; Scheromm 2013). They are characterized as one parcel of land exploited collectively by an ensemble of gardeners. However, the parcel is often divided into small plots at the request of the gardeners, so that each gardener then has the exclusive use of his own individual plot. These gardens were recognized for their role in social relationships, and their links to nature and the environment, during the first national forum of gardening and citizenship which took place in Lille in 1997. They were presented as examples of citizens reappropriating abandoned spaces, and contributing to a perspective of urban citizenship (Giband and Siino 2013). This forum marks the appearance in France of an image of shared gardens as providing a leading social role, and a symbol of citizens investing in urban space. As a more recent concept than that of family gardens, their identity is based on values of solidarity, conviviality, links and sharing between generations and cultures, as well as respect for the environment and the reintroduction of natural landscapes in the city. Their proliferation has reintroduced gardening as a fashionable urban practice, and encourages the evolution of practices in family gardens.

Contrary to the case in France, Portuguese collective gardens do not benefit from a legal definition, but the actors and scientists concerned with them use the generic term hortas urbanas, urban gardens (Castel'Branco et al. 1985; Pinto 2007; Folgosa 2014). This designation groups all forms of gardens which make use of urban land for the production of foodstuffs. The development of urban gardening in Portugal is

Montpellier. From left to right. Municipal family gardens. Inauguration of a municipal shared garden in the city's historic center. Informal shared garden created by a group of friends on the lands of a farmer in the north of the city.

Lisbon. On the left: social gardens with large plots. In the middle: leisure gardens with medium plots. The two first parks are planned by the municipality. On the right: informal gardens created spontaneously by the inhabitants close to their homes.

Fig. 12.1 The different types of collective gardens in Montpellier and Lisbon (Sources: pictures of Montpellier, Scheromm P. 2013; pictures of Lisbon, Mousselin G. 2013)

driven by two concomitant factors: the flight of rural populations to urban centers and the influx of immigrants from previous Portuguese colonies following their independence in 1975. The massive addition of new urban citizens with a tradition of cultivating the land was accompanied by an emergence of informal gardening at the same time that more traditional agricultural activities were disappearing in the wake of urban expansion. This informal agriculture is characterised by a spontaneous spatial occupation taking the form of anarchically grouped, small garden "squats" (Fig. 12.1). As a result of their contravening aspect, these clandestine gardens have long remained outside of urban regulations, at both the national and local level. Their emergence can be traced to the motivation of citizens responding to their basic needs for food. The dynamics in the development of these gardens has been reinforced by the crisis, and has alerted public authorities. Since the middle of the 2000s, several Portuguese cities have seen local authorities implement projects for planned gardens. The criteria for these planned gardens is usually defined on a municipal scale. In Lisbon, they are grouped in horticultural parks governed by strict regulations that focus on two primary types of gardens: social gardens, and leisure gardens. The organization of the gardens is relatively similar, and close to that of informal gardens. They are subject to rental regulations instituted by the municipality (Ramos 2011). The social garden is fundamentally designed to respond to the food production needs of the gardener, while the leisure garden is aimed towards an improvement in the quality of life. There is a significant difference in plot size between the two garden categories. The social gardens are larger in order to guarantee their function of food security, as are the parcels solicited for informal gardens. As in France, the planned gardens are fenced, equipped with sheds, and connected to a public water system. In contrast, the informal gardens tend to use recuperated materials to build fencing and sheds to store crops and tools.

Montpellier's gardens stem from two models: the model of nineteenth century "workers' gardens", and American community gardens. They represent a national and transnational dynamic. Conversely, the gardens of Lisbon draw more from local processes and rural practices transposed to an urban setting, and there are no national regulations framing their functions.

12.3.2 Characteristics of Collective Gardens in Montpellier and Lisbon

12.3.2.1 In Montpellier, Small Collective Gardens Cultivated by Any Urban Resident

Thirty collective gardens, family or shared, were identified in Montpellier. They occupy a surface area of approximately 11 hectares, primarily on public property (Scheromm et al. 2014).

The oldest existing gardens are family gardens created by enterprises, gardening associations, or the municipality. There are six of these gardens located on the periphery of the city (Fig. 12.2), and together, they comprise a group of up to 60 plots, each between 100 and 200 square meters.

Twenty-four shared gardens were identified. They are situated in different sections of Montpellier, including the city's historic centre. The oldest dates from 2005. Three of them are student gardens, installed on the campus of their school. The others belong to the municipality, with the exception of a garden created by an association for social integration, and an informal garden created by individuals that have been unsuccessful in obtaining a plot in the municipal family gardens. These gardens, typically small in size, are found in single parcels varying between 100 and 400 square meters.

Montpellier's collective gardens are cultivated by a socially heterogeneous representation of urban residents. The family gardens situated at some distance from the historic city centre are in neighbourhoods principally populated by employees, workers and mid-range professions. However, the plots in Montpellier's family gardens are allocated in a drawing, which means that the citizens cultivating in family gardens are not necessarily inhabitants of the neighbourhood where the garden is located. The shared gardens are located in neighbourhoods with contrasting socio-demographic profiles that include significant differences in the socio-professional categories represented among residents: the percentage of neighbourhood residents from upper professions requiring advanced degrees varies from 1% to 59%. Although some of these gardens are found in modest areas at some distance from

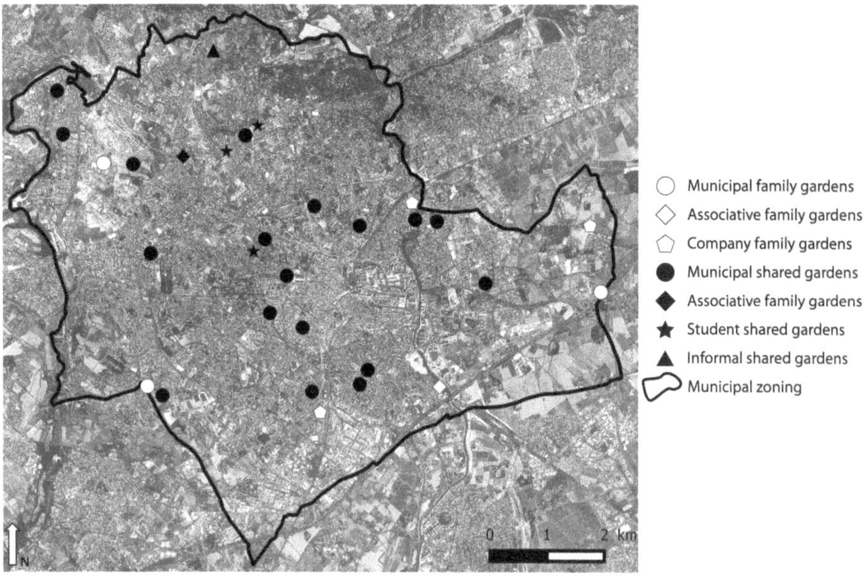

Fig. 12.2 Map locating collective gardens in the city of Montpellier (Sources: picture IGN BD Ortho 2006, inventory Scheromm P. 2013)

the city centre, as seen with the family gardens, others are in affluent areas situated in close proximity to the historic centre. These observations show that shared gardens develop in various types of neighbourhoods. The investigation conducted in the family and shared gardens confirms that gardening activities can be of interest to any urban resident, regardless of her/his socio-professional category (Scheromm 2015). The gardeners cultivate their plots from a perspective that reflects the pursuit of pleasure and occasionally a militant posture towards food production. The upper and intermediate socio-professional categories are present, but retired and mid-range employees represent a larger percentage of the urban citizens cultivating these gardens.

12.3.2.2 In Lisbon, Large Collective Gardens Cultivated by Poor Citizens

Lisbon's planned gardens currently occupy a surface area of approximately 14 hectares. According to city officials, approximately 400 families were participating in 11 parks at the end of 2014. These horticultural parks claim the fundamental distinction of social or leisure gardens. Although no data exist to support this contention, the type of park depends strongly on the socio-demographic characteristics of the area where it is located. The social horticultural parks group up to several 100 plots with a minimum size of 100 square meters, but the surface area of the plots in leisure gardens is much smaller. The number of plots available in each park correlates to the availability of municipal property rather than to the demand from urban

Table 12.1 Characteristics of the first planned horticultural parks

Park	Profile	Number of plots
Graça	Social	6
Quinta N. S. da Paz	Leisure	9
Bensaùde	Leisure	20
Teilheiras	Leisure	21
Jardins de Campolide	Leisure	22
Olivais	Social	31
Odivelas	Social	33
Quinta da Granja	Social	38
Chelas	Social	400[a]

Sources: Lisbon municipal master plan 2014; Gonçalves (2014)

[a]This number represents the plots anticipated for the entire project, but according to the latest information, the number will likely drop as the municipality supports other developments in the project

citizens. One social horticultural park, Chelas, encompasses several 100 plots of 150 m^2, while another social park, Graça, only holds six plots (Table 12.1). The city also has horticultural parks that have been developed through the direct initiatives of civil society actors. One emblematic example is the Alta de Lisboa park which was conceived in 2009 by an environmental association and inaugurated in 2014. This park is located on a parcel provided by the city, and includes 70 plots from 50 to 100 m^2 each, a 400 m^2 cultural space with handicap access, as well as another 2 hectares designated for the development of other agricultural activities.

By contrast, it is very difficult to know the precise number of informal gardens. They take place here or there, on unoccupied open spaces, or vacant spaces waiting to be developed (Fig. 12.3). In a 2009 municipal census, they represented 84 hectares scattered throughout the city. The largest pockets of informal cultivation lie on the periphery of the city where open spaces are more abundant, although a few small examples exist in the historic centre. The presence of these informal gardens correlates strongly to the social, cultural, and economic context of the neighbourhoods where they develop. In the majority of cases, they constitute an indicator of more marginalized zones where people live under precarious conditions. Reflecting these conditions, the gardens respond to the needs of families in the sense of a food supply and an economic opportunity. Actually, a portion of the harvest is often sold, and because these gardens often develop on land that may be polluted, this commercialization of foodstuffs creates problems in sanitary control.

In both cities, the urban collective gardens are usually developed on municipal property. They encompass considerably more surface area in Lisbon than in Montpellier. The family gardens in Montpellier are similar in form to the planned gardens in Lisbon. However, we observe differences between the two cities. Informal gardens are found on dozens of hectares in Lisbon, where gardeners from the most economically and socially vulnerable population cultivate primarily for

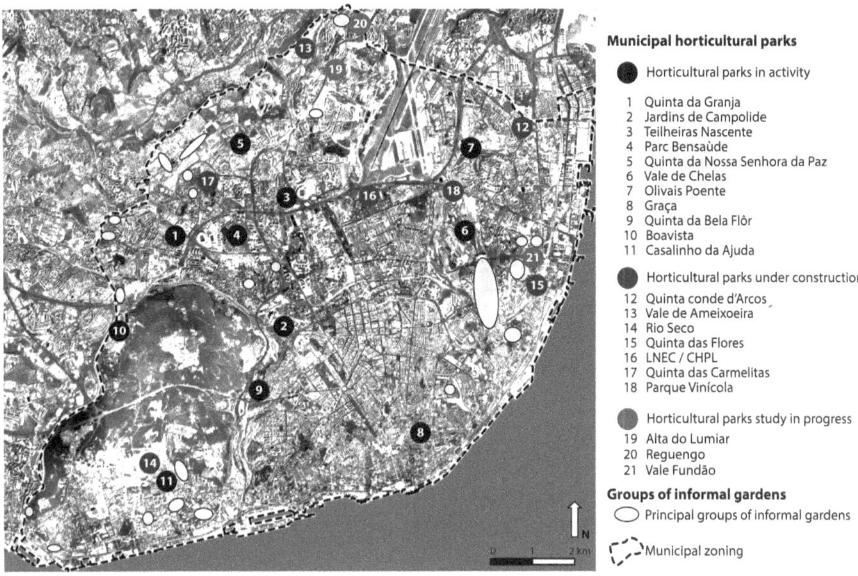

Fig. 12.3 Map locating informal gardens and horticultural parks planned by the municipality of Lisbon (Sources: Lisbon municipal master plan 2012 and municipal data 2014; inventory Mousselin G. 2012–2014)

food production. In Montpellier, collective gardens are more likely to represent small spaces for leisure in a context of nature and social interaction, and the participating urban gardeners reflect more diverse social profiles.

After identifying and locating these collective gardens, we examine the dynamics involved in their creation and functions.

12.4 The Recent Proliferation of Collective Gardens: Urban Residents Demand and Municipal Support

12.4.1 In Montpellier

Three of Montpellier's family gardens date from the 1970s. They were developed through civil society initiatives and are managed respectively by a national association (collective gardens of the French National Railway Company), a local association (the Gardeners of the Hérault), and by an enterprise committee (EDF gardens). Three additional family gardens have been created by the municipality since 2004, and are managed by the city's green space department. Discussions with city officials reveal that their creation was motivated by persistent demand from citizens who have expressed dissatisfaction over the course of several years with the lack of gardening space in the city.

The first shared garden in Montpellier was organized in 2005 as a response to the mobilized effort by residents of a central neighbourhood to preserve their quality of life. The effort resulted in a combined urban park and shared garden created as an alternative to replace a contested project for the construction of 200 residential units. Since this time, there has been a steady increase in the development of shared gardens, always at the demand of citizens or associations (Mousselin and Scheromm 2015). In this process, we see residents, often organized in associations or neighbourhood committees, working to secure available property from the municipality, which then provides logistical and financial support for the creation of a shared garden. Other institutional actors are also involved in their creation. The Roseraie Sainte Odile garden was developed on the grounds of a residence for the elderly with the aid of the municipality. In the spirit of social integration, this garden project is to open to neighbourhood residents as an effort to encourage intergenerational exchanges. Jupiter square is a shared garden created in the interior courtyard of a social housing complex in a disadvantaged neighbourhood. The project was initiated by an association of social and cultural intervention, and again, with municipal support and collaboration. This courtyard acted as a virtual dumping ground before the creation of the garden. The appropriation of the space by gardening residents has reduced the accumulation of garbage thrown from the buildings' windows, and created a striking green space in the midst of an otherwise concrete jungle.

Most of Montpellier's shared gardens operate in a tripartite fashion: the municipality participating through the office of local culture (which also registers memberships); the neighbourhood committee or association that initiated the project; and an association appointed by the municipality to insure the garden's organization and the development of gardening programs called "Green Hand". A project manager of the city's green space department takes care of specific maintenance and management of the gardens. This ensures a strong link between the gardeners and the municipality.

The municipality has a major role in the proliferation of both shared and family gardens in the city. It provides for most of the network's management, and contributes to social and pedagogic functions by financing gardening programs presented by social, nature, and environmental associations. But the municipality's central role in the proliferation of collective gardens is ultimately a result of the demand from urban residents. It is this demand that drives the development and existence of these gardens. There is a relatively diverse number of civil society actors that also participate in the development of collective gardens. They are primarily interested in the social aspects of the projects or the creation of green space as it relates to the quality of life. The governance of the most recent gardens is clearly cantered around the municipal actors, formal groups of citizens (committees or associations), and organizations operating in social and environmental fields (Fig. 12.4).

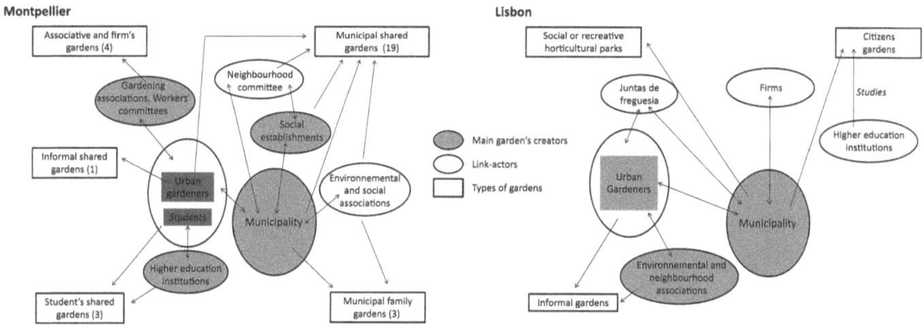

Fig. 12.4 Actors system of collective gardens in Montpellier and Lisbon. One-way arrows connect gardens with the actors involved in their creation or management. Two-way arrows identify relationships between the different actors involved

12.4.2 *In Lisbon*

Urban residents are very active in the creation of urban gardens, particularly among disadvantaged populations such as undocumented immigrants, and the municipality plays an important role in their development and management. City officials have long acknowledged the existence of informal gardens, ultimately adopting a posture of tolerance, and even going so far as encouraging and legitimizing their installation through procedures ceding or leasing land. The creation of horticultural parks, however, is a more recent phenomenon. These planned gardens are conceived as a response to the needs of residents in the various neighbourhoods of the city. The municipality remains the owner of the land on which the park is developed, and ensures the management of communal portions of the park by the city's green space department. The city's social services can be involved indirectly through a framework of social measures targeting marginalized neighbourhoods. Plans for the first horticultural parks were unveiled in 2009 for zones identified as strategic for addressing the needs of urban residents, and supporting the development of urban green space. All of these parks are organized according to the same management structure, and adhere to the same regulations in the selection of applicants – the allocation of plots is conducted through a competition. Unlike Montpellier, the allocation process for garden plots in Lisbon is designed to give preference to prospective gardeners that live near the garden. This selection policy is designed specifically to limit commuting and facilitate the accessibility to each park. In addition, it encourages interactions among residents that create social links on a neighbourhood scale. The municipality regularly publicizes future parks and calls for applications; residents can also find information at one of the municipal reception offices that distribute applications and ensure their reception. These horticultural parks can be created in new areas, or even as replacements for existing informal gardens. In the latter case, the municipality attempts to accommodate former occupants of the space by reserving plots for them in the organization of the new park. The Juntas de freguesia (Portuguese district administrative unit) can use its credentials as an integral part of the neighbourhood to act as an intermediary, and is sometimes called upon to intervene in conflicts involving gardeners.

The municipality has developed a logic for advancing urban agriculture based on centralized governance (Fig. 12.4). Nevertheless, some of the gardens are the result of initiatives launched by residents and associations, and in these cases the municipality can act as a partner. The Alta de Lisboa park, which was noted earlier in this chapter, is an example. Commercial enterprises or foundations also can be involved in financing horticultural parks or smaller collective gardens. Vitacress is a company that has participated in the financing for the municipal park, Teilheiras, and the EDP Foundation (created by the Portuguese energy company, EDP) participates through the framework of the project, Horta Solidária (Solidarity Garden), designed to support the creation of parks or collective gardens. Studies are being conducted by research groups in order to better understand the social and environmental benefits of these gardens and the ways in which they enhance the quality of life for urban residents. The existence of programs and events (meals, tours, contests, etc.) organized by the gardeners themselves in many of the parks, reveals the strong social links forged through the activity of gardening.

In Montpellier as in Lisbon, the popularity of collective gardens reflects the demand and the needs of citizens. A variety of civil society actors have been mobilized in the efforts to create these gardens: urban residents themselves, associations, neighbourhood committees, students in collaboration with their schools, spontaneous citizen groups. The municipality in turn complements this demand by financing the development of collective gardens and coordinating their functions. In Lisbon, poor residents are the direct instigators of numerous informal gardens but are not implicated in the governance of the gardens. In contrast, the Montpellier's collective gardens exhibit public participation in their governance, particularly in decisions concerning the development of shared gardens.

12.5 Institutional Recognition and Development of a Dedicated Public Policy

12.5.1 In Montpellier

The municipality plays a very important role in the collective gardens of Montpellier. Only five of the gardens identified have little or no collaborative relationship with city authorities. For the all others, the city has provided municipal property, financing, management, and development support, in close cooperation with neighbourhood committees or associations from the environmental or social fields. By promoting citizen participation, this style of governance has enhanced the relationships between actors involved in the issues surrounding the development of these gardens, and between gardeners and the elected officials charged with the management of green space in the city. This collaboration contributes to the creation of an urban culture of collective gardening. Gardeners are also being sensitized to the advantages of organic gardening methods. The municipal charters granting the use of these parcels strongly recommend following the practices of organic cultivation, and numerous workshops are presented to educate gardeners about organic methods. This sensitization leads to agroecological practices, which in turn add another dimension to the social and leisure aspects of the collective garden (Scheromm 2015). Collective gardens are periodically open to the public for educational tours which contribute to the general awareness of urban gardening. The coordinated efforts of various participants suggest that these gardens form a social and recreational system of horticulture in the city. However, the public policies concerning these gardens are still incomplete. Even though their development is included in the planning strategy preserving biodiversity (Scheromm et al. 2014), the collective gardens are not specifically mentioned in Montpellier's master plan, and their creation is relegated to periodic episodes based on the demand from urban residents (Mousselin and Scheromm 2015). They are not the object of a dedicated local public policy, so the recent trend of proliferation could end at any moment. Nevertheless, Montpellier's urbanism project for 2040, and the most recent food and agricultural policy for the metropolitan region, do place strategic importance on urban agriculture. These elements combined suggest that a more inclusive policy for different types of agriculture in the city could be embraced in the years to come.

12.5.2 In Lisbon

Lisbon's horticultural parks are imbued with a very strong municipal presence, much like that found in Montpellier. The municipal regulations for these parks also advocate organic methods in cultivation, and there are programs in place, designed to sensitize gardeners and the general public to the advantages of organic production. Nevertheless, the gardeners themselves have very little direct contact with the city's central authorities. Problems and conflicts are usually resolved by the city's neighbourhood representatives.

The number of collective gardens in the city are increasing. There are 11 existing parks and 10 more are projected. Seven of the latter are currently under construction and three are still in a study phase. Eventually the municipality plans to develop local markets linked to the horticultural parks as a means to highlight and distribute intra-urban agricultural production. In 2014, the municipality created a wine-making park in conjunction with new gardens on 3 hectares near Lisbon's international airport. This political action represents a new stage in the reintroduction of agriculture within the city.

The inclusion of these parks and urban gardens in the municipal's urban regulations is also evidence of a political will for these initiatives. As in France, the regulations presented in the Lisbon, municipal master plan are based on a strict nomenclature that defines land use and development possibilities. The addition of collective gardens into urban zoning regulations assures additional support for the urban cultivation movement in the Portuguese capital. The development of a municipal agricultural strategy is a strategic axis in the response to the various intersecting issues of sustainable urban development. Additionally, beyond their production capacity, these parks and gardens are seen as green strongholds that participate in a more ecological plan for the city.

The municipalities of Montpellier and Lisbon have emerged as key players in the development of collective urban gardens, particularly through the encouragement of environmentally responsible gardening. Both cities have established a process that truly institutionalizes these gardens. This process exerts a stronger presence in Montpellier because the municipality is at the origin of most of the gardens. But this activism has not yet translated into official public policy on a local scale; the creation of collective gardens is not among the objectives of official urban planning. In Lisbon, there is still a significant number of informal gardens. In this sense, the institutionalization of collective gardens is not as comprehensive as that observed in Montpellier. However, Lisbon's formal gardens are more integrated into public policy, as evidenced by their inclusion in official provisions for the organization and management of open spaces. For the gardeners themselves, the function of collective gardens in Lisbon is, above all, one of food production, but the municipality also acknowledges their function in a framework of the quality of life and the addition of urban green space. Along with other activities, they contribute to "green" the city.

12.6 Conclusion

The collective gardens of Lisbon and Montpellier are characterized by a combination of commonalities and differences.

Indicative of the needs of the urban residents, the proliferation of urban collective gardens in the both cities began in the first decade of this century. In both cases the demand from urban residents is at the origin of their development, whether that demand arises out of economic or social difficulties, or the need to enrich the urban lifestyle that otherwise distances man from his natural environment. In the course of their evolution, the urban collective gardens have become an important element in the quality of urban life from the perspective of both urban residents and the municipalities, who contribute to their legitimate place in the urban landscape. The authority over available land gives municipal actors a privileged position in the development and management of these gardens. They become the undisputable powerbrokers in the establishment of gardens on the available urban interstices or city perimeters. The urban citizens also play a major role in the development of collective gardens, either directly through their creation as evidenced by Lisbon's informal gardens, or as facilitators in various parts of the process. Associations and citizen groups or neighbourhood committees can indeed be seen in the process of motivating and supporting the development of formal and informal gardens. A part of the collective gardens can be considered as institutionalized forms of gardening, supported by the political will of the municipality. But their sustainability as an element of urban landscapes is not guaranteed, except perhaps in the case of Montpellier's family gardens that are protected by law, or Lisbon's horticultural gardens that are written into the provisions for open space in the city. The ability of other garden forms to survive will have to pass the test of time. The illegal character of Lisbon's informal gardens mainly leaves them vulnerable in the face of other development and infrastructure priorities.

The two municipalities organize the size of individual garden plots on a similar scale, but the gardens in Lisbon hold a more important spatial presence (84 hectares) than those in Montpellier (11 hectares). This important difference must be placed in the respective contexts. In Lisbon, the existence of numerous informal gardens illustrates the precarious socio-economic situation of residents, particularly in outlying neighbourhoods and certain areas of the city centre. These gardens are cultivated for their production of foodstuffs, either for the gardener's own consumption or for sale as a source of revenue. If the municipality may associate collective gardens with social and environmental issues, but the gardeners themselves are mainly interested in the security and economics of food production. This gardening activity can be question of survival for some households in extremely precarious positions. Conversely however, the gardeners of Montpellier primarily identify with the social, environmental, and leisure aspects of their gardens. They do not garden to ensure the security of their food supply, even if their garden represents a desire to take more personal control over the food available for their consummation (Brand and Bonnefoy 2011; Scheromm 2015).

The comparison of the two cities underlines the articulation between the citizens demand and needs; it also underlines the multifunctionality of collective gardens. Their functions are specific to the context within which they develop. It is also connected to the different regulations and legal status under which the gardens are cultivated: the gardens in Montpellier are regulated by a national law which prohibits the sale of the garden's production; while the sale of production from the gardens in Lisbon is authorized by municipal policies which envision plans for an organized urban market. The comparison also points to the simultaneous institutionalization process occurring in both cities, as well as identifying the specificities of this process in each case. The proliferation of collective gardens in Montpellier is the reflection of a municipal response to the demand by urban residents. In Lisbon, it has been driven by the appropriation of vacant spaces by an impoverished population and the municipality's political commitment to regularize this phenomenon and develop it.

As tools that are multifunctional and extremely popular with urban residents, the creation of collective gardens has surged during various historical periods of crisis. Today, their forms are evolving through new perspectives on the quality of urban life and the urban landscape (Blanc 2010; Bourdeau-Lepage and Vidal 2013). The issues surrounding their place in contemporary cities highlight ideological and political concerns: How can we establish agriculture as an integral component in the quality of urban life, one that serves urban residents through agricultural production, and also responds to their needs for a natural environment and outlets for social interaction? The recent resurgence of collective gardens in Montpellier and Lisbon, and their institutionalization through municipal public policies, adhere to this logic in the construction of a model for sustainable cities.

Acknowledgements This research was financed by the French National Research Agency: DAUME project n° ANR-2010-STRA-007-01.

References

Blanc N (2010) L'habitabilité urbaine. In: Coutard O, Levy J-P (eds) Écologies urbaines. Economica, Paris, pp 169–189

Bourdeau-Lepage L, Vidal R (2013) Nature urbaine en débat: à quelle demande sociale répond la nature en ville ? (pp. 195–210). Demeter 2013. http://www.clubdemeter.com/ledemeter.php?demeter=2013

Brand C, Bonnefoy S (2011) L'alimentation des sociétés urbaines: une cure de jouvence pour l'agriculture des territoires métropolitains ? VertigO – la revue électronique en sciences de l'environnement 11(2). http://vertigo.revues.org/11199. https://doi.org/10.4000/vertigo.11199

Cabedoce C, Pierson P (1996) Cent ans d'histoire des jardins ouvriers. Creaphis, Grane

Castel'Branco I, Saraiva M, Neto MS (1985) As "hortas urbanas" em Lisboa. Sociedade e Território 3:100–107

Consales, J-N (2004) Les jardins familiaux à Marseille, Gênes et Barcelone: Laboratoires territoriaux de l'agriculture urbaine dans l'Arc Méditerranéen. Doctoral thesis in Geography. University of Aix Marseille 1

Dubost F (1997) Les jardins ordinaires. L'Harmattan, Paris

Duchemin E (2013) Multifonctionnalité de l'agriculture urbaine : Perspective de chercheurs et de jardiniers. In: Duchemin E (ed) Agriculture urbaine : Aménager et nourrir la ville. VertigO, Montreal, pp 97–107

Duchemin E, Wegmuller F, Legault M (2010) Agriculture urbaine: un outil multidimensionnel pour le développement des quartiers. VertigO – la revue électronique en sciences de l'environnement 10 (2). http://vertigo.revues.org/10436, https://doi.org/10.4000/vertigo.10436

Ernwein M (2014) Framing urban gardening and agriculture: on space, scale and the public. Geoforum 56:77–86. https://doi.org/10.1016/j.geoforum.2014.06.016

Folgosa R (2014) Parques hortícolas municipais: Uma estratégia para a Estrutura Verde de Lisboa. http://www.adrave.pt/uploads/writer_file/document/510/Apresenta_o_Rita_Folgosa.pdf

Giband D, Siino C (2013) La citoyenneté urbaine pour penser les transformations de la ville ? Annales de Géographie 694(6):644–661. https://doi.org/10.3917/ag.694.0644

Gonçalves RGG (2014) Hortas urbanas: Estudo de caso de Lisboa. Agronomic engineering thesis. Lisbon University

Guyon F (2005) Les jardins familiaux : miroirs des politiques de la cité. Loisir et société 27:529–546

Hallsworth I, Wong A (2015) Urban gardening realities: the example case study of Portsmouth, England. Int J Food Syst Dyn 6(1):1–11. 10.18461/ijfsd.v6i1.611. http://centmapress.ilb.uni-bonn.de/ojs/index.php/fsd/article/view/437/433

Hardman M, Larkham PJ (2014) The rise of the "food charter": a mechanism to increase urban agriculture. Land Use Policy 39:400–402. https://doi.org/10.1016/j.landusepol.2014.02.022942

Huang D, Drescher M (2015) Urban crops and livestock: the experiences, challenges, and opportunities of planning for urban agriculture in two Canadian provinces. Land Use Policy 43:1–14. https://doi.org/10.1016/j.landusepol.2014.10.011

Mansfield B, Mendes W (2013) Municipal food strategies and integrated approaches to urban agriculture: exploring three cases from the global north. Int Plan Stud 18(1):37–60. https://doi.org/10.1080/13563475.2013.750942

Mousselin G, Scheromm P (2015) Vers une mise en politique des jardins collectifs urbains: approche comparée de deux trajectoires municipales à Montpellier et à Lisbonne. J Urban Res 6. http://articulo.revues.org/2685, https://doi.org/10.4000/articulo.2685

Pashchenko O, Consales JN (2010) Les jardins collectifs: derrière une seule notion, des réalités territoriales contrastées, Durabilis, magazine expérientiel de l'environnement et du développement durable. http://www.magazine-durabilis.com/les-jardins-collectifs-des-realites-territoriales-contrastees/4/

Pinto R 2007 Hortas urbanas: Espaços para o desenvolvimento sustentável de Braga. Engineering thesis. University of Minho

Pourias J, Duchemin E, Aubry C (2015) Products from urban collective gardens: food for thought or for consumption? Insights from Paris and Montreal. J Agric Food Syst Community Dev 5(2):175–199

Pudup MB (2008) It takes a garden: cultivating citizen subjects in organized garden projects. Geoforum 39(3):1228–1240. https://doi.org/10.1016/j.geoforum.2007.06.012

Ramos, AR (2011) A integração de espaços de cultivo agrícolas em contextos urbanos: Proposta de intervenção para a requalificação urbana do vale de Chelas (Lisboa). Architecture thesis. Higher Technical Institute, Lisbon University

Reyburn S (2002) Le cadre de vie et les jardins potagers communautaires à Montreal. VertigO – la revue électronique en sciences de l'environnement 12(2). http://vertigo.revues.org/3794. https://doi.org/10.4000/vertigo.3794

Scheromm P (2013) Les jardins collectifs, entre nature et agriculture. Métropolitiques. http://www.metropolitiques.eu/Les-jardins-collectifs-entre.html

Scheromm P (2015) L'expérience agricole des citadins dans les jardins collectifs urbains. Le cas de Montpellier. Développement Durable et Territoires 6(1). https://developpementdurable.revues.org/10726. https://doi.org/10.4000/developpementdurable.10726

Scheromm P, Perrin C, Soulard C (2014) Cultiver en ville... Cultiver la ville ? L'agriculture urbaine à Montpellier. Espaces et Sociétés 158:49–66

Steel C (2009) Hungry city: how food shapes our lives. Vintage, London

Telles RG (1997) Plano Verde de Lisboa: componente do Plano Director Municipal de Lisboa. Colibri, Lisbon

Vitiello D, Wolf-Powers L (2014) Growing food to grow cities? The potential of agriculture for economic and community development in the urban United States. Community Dev J 49(4):508–523. https://doi.org/10.1093/cdj/bst087

Chapter 13
The Governance of Urban Agriculture and Multifunctional Land Use in the City of Zurich

Ingrid Jahrl and Otto Schmid

Abstract Aspects of multifunctional land use are increasingly gaining the attention of different urban actors in the city of Zurich, including urban gardeners, farmers and policy-makers as well as city administrators. This contribution is based on an exploratory case study on approaches and challenges for multifunctional land use with regard to urban agriculture in Zurich, including personal interviews, document analysis and a workshop with different actor groups. The city of Zurich, as the owner of a large area of farmland, has taken several approaches to multifunctional urban agriculture, which follow typical aspects of rural development, mainly targeting farmers. City dwellers, who are active in urban agriculture and gardening initiatives, increasingly challenge the current policies. The study results showed that there are different interests for land use between, but also within, the various urban actor groups. In particular, farmers partly consider the diverse land use functions as conflicting objectives within the multifunctional concept. Varying actors with similar land use interests may compete for control of land due to their different forms of organisation such as professional urban gardening initiatives and farmers, as well as more hobby oriented gardening initiatives and allotment gardeners. This new and potentially conflicting situation illustrates the need for an integrated urban approach of land use planning and territorial governance towards multifunctional land use in cities; which bridges interests and fosters co-operation between different actors (city administration, farmers, gardeners in urban gardening initiatives, traditional allotment gardeners).

Keywords Multifunctionality · Urban agriculture · Urban gardening · Land use policy

I. Jahrl (✉) · O. Schmid
Research Institute of Organic Agriculture (FiBL), Frick, Switzerland
e-mail: ingrid.jahrl@fibl.org; otto.schmid@fibl.org

C.-T. Soulard et al. (eds.), *Toward Sustainable Relations Between Agriculture and the City*, Urban Agriculture, https://doi.org/10.1007/978-3-319-71037-2_13

13.1 Introduction

As societies become more urbanised, politicians and planners increasingly face competing demands for the use of scarce land in and around cities. Multifunctional land use, which combines different functions within one area is offered as a solution to overcome this dilemma (Deelstra et al. 2001; Paracchini et al. 2011). In this regard, urban agriculture offers an alternative land use concept, which integrates multiple functions in areas which are densely populated (Lovell 2010; Wiggering et al. 2006). Urban actors, planners and policy actors are gaining in awareness of these ideas and are becoming increasingly interested in agriculture (Aubry et al. 2012; Zasada 2011) and in the issue of food and food production (Morgan and Sonnino 2010). Agricultural areas within cities are no longer seen only as land reserved for future urbanisation, but rather as a possibility to foster sustainable development of cities (Aubry et al. 2012). Urban agriculture is recognised as contributing to the sustainability of cities from a social, economic and environmental perspective (Deelstra and Girardet 2000). It tackles issues related to regional food production and food security (Drescher 2001). And it has the potential to improve the microclimate, minimise waste, improve nutrient recycling and water management, encourage biodiversity and enhance the environmental awareness of city inhabitants (Deelstra and Girardet 2000). Furthermore, urban agriculture can provide environmental amenities, accessible green open spaces and recreational services. Farmers in and around cities increasingly engage in landscape management and agri-environmentally friendly production, social farming or recreation-oriented diversification (e.g. farm based tourism). However, farmers are not the only people cultivating land in cities. Changes in lifestyle, a 'quality of life' orientation and growing concerns about the environment and climate change contribute to urban society's increasing interest in having agriculture at its doorstep (Zasada 2011). This leads to an increasing interest of civil society in aspects of food production, which might result in active involvement in urban agriculture (Renting et al. 2012). So-called new forms of urban agriculture (La Rosa et al. 2014) include a diverse range of ways in which consumers may become active in food production such as community supported agriculture (CSA) or collective urban gardening initiatives.

While the potential environmental, social and economic benefits provided by diverse forms of urban and peri-urban agriculture regarding multifunctional land use have been highlighted (La Rosa et al. 2014; Barthel and Isendahl 2013), scholars have raised critique on the concept of agricultural multifunctionality, seeing it as a rather theoretical concept and/or as a way to justify continued protection of the agricultural sector (Schmitz and Moss 2005). In practice, the concept is confronted with different interests and conflicts for land use in terms of functions the land should fulfil and what the land in rural and (peri-)urban areas should be used for (Torre and Darly 2014; von der Dunk et al. 2011; Mann and Jeanneaux 2009). The pursuit of multifunctional solutions and the new forms of urban agriculture challenge traditional governance mechanisms of agri-food systems and urban planning (Selman 2009; Renting et al. 2012). Challenges in this concern are seen out of the increased interest of private and semi-public actors with heterogeneous preferences for a greater involvement in decision-making processes and local projects. Furthermore, actors are confronted with different location-based dimensions of policy-making, the local, regional, national and the European level with its trail of decisions and regulations (Torre and Traversac 2011).

This chapter contributes to debates on territorial governance of urban agriculture, providing a better understanding of local actors' perspectives (city administration, farmers, urban gardeners) on multifunctional land use of urban agriculture. For this, we explore the dynamics of multifunctional land use in the city of Zurich by asking: What are the perspectives of different actors towards agricultural land use in an urban area? What governance approaches and challenges can be identified in shaping multifunctional land use of urban agriculture?

13.2 Methods

This research followed an exploratory case study approach. Exploratory studies are mainly conducted for investigating an issue or a topic aiming to develop insights and ideas about its underlying nature (McNabb 2010). This approach was utilised to gain an understanding of governance mechanisms concerning multifunctional land use for the case of the city of Zurich. Zurich was chosen as a case study as city policy has put in place different instruments for governing multifunctional land use, which to our knowledge, differs from many other European cities.

Applying a case study approach, which 'investigates a contemporary phenomenon within its real-life context; when the boundaries between phenomenon and context are not clearly evident; and in which multiple sources of evidence are used' (Yin 1989), allowed us to systematically collect and describe the perspectives of different actor groups. Actor groups were in particular public administration, market actors, civil society organisations and individuals, following a classification from Wiskerke (2009). Key messages were gained through qualitative interviews and document analysis.

Semi-structured interviews were conducted with 18 participants: five key actors from the city administration, six farmers (who farm city owned land) and farmer representatives, in addition to seven actors of civil society, such as representatives and individual gardeners of urban gardening initiatives and allotment gardening associations. The city administration participants included representatives related to sustainable public food procurement ("Department of environmental protection") and three sub-divisions of a department which deals with different aspects of multifunctional land use (land planning, nature protection and education, agriculture and forestry, with the "Department of green space management").

The actors were chosen with the objective of identifying different perspectives within these three main groups. The interviews focused on the actors' approaches and perspectives towards multifunctional land use of urban agriculture. Further insights were gained through a workshop with 14 participants from the three actor groups mentioned above; most of whom who had been interviewed. The workshop focused on the actors' perspectives, ideas and challenges concerning the overall development of a sustainable food system in the city. The interviews and workshop data were transcribed and analysed using a qualitative content analysis approach (Mayring 2002), applying an inductive category development and deductive category application for text interpretation focusing on the different views on the topic held by the actors. The results from this analysis were synthesised and reported (Schmid and Jahrl 2014).

13.3 Agriculture in Zurich City

Zurich is the largest city in Switzerland, with approximately 400,000 of the 1.83 million inhabitants of the Zurich metropolitan area living within the city boundary. The city is located in the northeast of Switzerland, in the canton Zurich, and has an area of 91.9 km^2 with a high density of settlements, industry and roads. Zurich is an economically booming region in Switzerland (Stadt Zürich 2013), which means there is strong pressure to build on agricultural land. However, 810 ha of agricultural land remains within the city, which contribute with forests and parks to a total green space area of 4440 ha (Fig. 13.1). In 2012 there were 25 full and 32 part-time farmers who cultivated agricultural land within (and also outside) the city border. The majority of the farms were mixed farms with animals, mostly cows or beef cattle but also laying hens, combined with arable and grassland. Full-time farmers cultivated between 20 and 170 ha, with an average of 30 to 35 ha. Ten of the 25 farms are owned by the city, which account for two thirds of the agricultural land. Nine of these farms are rented to family farmers, while one is run under the management of the city department in charge of green space management. The rented farms to family farmers on city land are farmed organically, while farms on private land are mainly managed conventionally or according to the guidelines of integrated production. Farms on city owned land tend to be a little bit bigger than private farms and have a slightly greater focus on on-farm processing and direct marketing as e.g. on-farm shops and direct sales of eggs to restaurants (Grün Stadt Zürich 2015a, b).

There is a long-standing tradition of allotment gardening, with about 5500 allotments totalling 135 ha. Allotment plots on city premises are administrated through 13 "family garden associations" who lease the plots to interested gardeners, whilst the city administration directly rents approximately a further 600 plots to gardeners. New urban gardening initiatives have arisen in the last years, and currently, 20 community gardens, migrant gardens or hobby animal holdings (sheep, bees) have been established on 2.8 ha of city owned land. One of these is organised as community supported agriculture (CSA) system with a vegetable box scheme and around 250 consumers. Furthermore, there are around 20 to 30 temporary community gardens; mainly on areas already allocated for construction, and where garden produce is mainly grown in boxes. Plots for urban gardening initiatives or bee keepers are

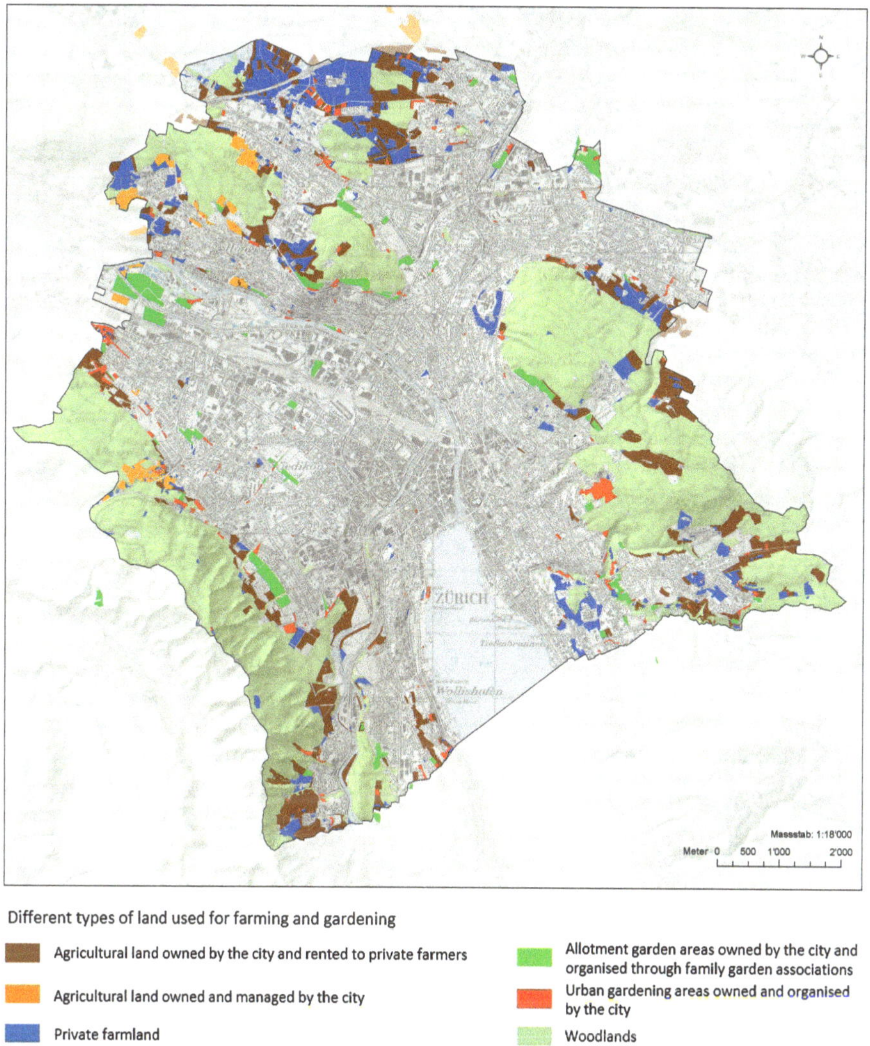

Fig. 13.1 Land for farming and gardening within the city of Zurich (July 2007) (Source: Grün Stadt Zürich, personal communication)

administrated directly through the city administration for green space management (Grün Stadt Zürich 2015a, b).

The policy and regulatory framework for the Zurich city is quite complex, comprising a mix of national, cantonal and city/municipality related laws, regulations and guidelines for implementation to meet varying policy goals. Zurich adopted the vision of becoming a sustainable city-region by the year 2025 and has defined a range of different sustainability goals that should be achieved by then (Stadt Zürich 2007). These goals especially target construction, mobility, infrastructure/traffic, energy use and public spaces. In 2008, the citizens of Zurich also voted for the city to adopt the goal of becoming a "2000-Watt-Society" by the year 2050, which would mean reducing energy consumption and CO_2 emissions, and supporting renewable energy sources. Agriculture and food do not play a significant role in the formulation of the "sustainable city-region" or the "2000-Watt-Society" activities, with only a vague description in the vision for 2025 that "in the year 2025, Zurich is an attractive city with an attractive landscape which is carefully cultivated".

Based on these broad sustainability goals, the city department for green space management has formulated strategic goals for its agriculture. Furthermore, a concept of Zurich city agriculture up to 2030 was developed and discussed in 2014–2015 within different divisions of the department for green space management. One of the department's strengths is that sections within the department, such as agriculture, nature protection, and land use planning can develop new strategies and concepts in close collaboration. The process continued on from a "Green Book" of the City Council, which included strategic goals in terms of agriculture from 2006 to 2016 (Grün Stadt Zürich 2015a, b).

13.4 Land Use Approach by Zurich City Policy and Administration

A strategy has been pursued in Zurich to condense settlements by e.g. constructing higher buildings with increased numbers of floors rather than expanding the built area into the surroundings. Existing urban green spaces are considered to be secured within the current long term land planning, which was formally outlined in the guiding master plan (in German "Richtplan") of Zurich. This plan is one of the central land use planning instruments for both the municipality and the canton of Zurich for future spatial development (settlements, agriculture, leisure areas, industry etc.). It places high importance on agricultural areas/zones, a strategy confirmed via a public vote in 2012, which voted for the protection and maintenance of valuable agricultural land. The legal barriers to development of green spaces in the city are much more stringent than in the past, but pressure from different interests within the city to release valuable green space land for construction increases. A trend towards urban sprawl can be observed as the city population increases and demands new infrastructure, such as new schools and sport areas, which are often constructed on allotment garden zones. The different interests in the city are coordinated between city departments through the City Council ("Stadtrat") and the department in charge of green space management in Zurich is eager to secure open green spaces by buying land from private or other public owners. This policy began in the 1930s when land as well as farms were bought by the city. In 2015, two thirds of the agricultural land belonged to the municipality of Zurich (Grün Stadt Zürich 2015a), which farmers on city-owned land interpret as evidence that the city has made a clear commitment towards agriculture. They are keen to continue agricultural production, and in this way also contribute to the maintenance of green spaces.

An empirical analysis of the city policies reveals that the agricultural land policy aims to serve different goals. Interviews with representatives of city administration reveals the following main tasks of agriculture in the city of Zurich: (i) designing and maintaining an attractive cultural landscape with high recreational value; (ii) preservation and promotion of biodiversity; (iii) production of food; and (iv) participation of the population and education. The first three multifunctional goals follow national agricultural policy in which multifunctionality is an important concept and includes goals such as food security, guaranteeing animal welfare standards, maintenance of cultural landscapes, ensuring that decentralised regions remain inhabited as well as nature conservation in terms of enhancing biodiversity (BLW 2015a). The last task of "participation and education" is specific to urban agriculture.

To fulfil these goals, the city of Zurich has taken different approaches towards farmed land within the city. The main pillars are: (1) binding requirements for organic farming practices for city-owned farms and urban gardens; (2) providing support and advice for agri-ecological measures on farms; (3) providing support for farm construction; and (4) fostering environmental education activities ("green knowledge") with city dwellers such as supporting school excursions to farms. These pillars are laid down in a strategy for the city agriculture, which is periodically revised by the City Council (Stadtrat Zürich 1999). They are also reflected in the informal criteria for tenant farmers, such as experiences in organic farming, being innovative, willingness to promote measures to improve biodiversity and being able to operate a farm in an urban context (communication with citizens, certain tolerance towards littering).

Organic cultivation of urban agricultural land was first introduced in the 1989 version of the agricultural concept for the city, and has been implemented in stages (pillar 1). The city department for green space management reports that more than 50% of all agricultural land within the city region of Zurich is organically operated (Grün Stadt Zürich 2015a). Following this decision, allotment gardens and urban gardening initiatives have also been obliged to be cultivated according to organic farming management practises. However, these obligations are difficult for the city administration to control and there is little additional support, such as courses, provided directly by city administration.

Agri-ecological measures on city farms are mainly encouraged by national policy measures (pillar 2), and to receive subsidy payments within the Swiss agri-environmental support scheme farmers are obliged to maintain 7% of their agricultural land as "biodiversity promotion areas", and fulfil other ecological minimum requirements (e.g. appropriate nutrient balance) (BLW 2015a; BAFU 2009). The city of Zurich has a higher obligation in place for its own farms, in which a minimum of 15% of farmland within the city must be dedicated to biodiversity promotion areas (30% was achieved in 2014). This local policy aims to counteract the depletion of landscapes and species decline in farming areas and contributes to a more attractive environment for citizens. City farmers can make use of national "biodiversity quality payments" (e.g. for species-rich meadows with a minimum number of indicator plant species) and payments for interconnectedness of ecologically-diversified areas, which are also linked to mandatory "landscape development projects" at the municipality level. These national payments are both financially attractive and supportive for city policies to diversify land-use systems for the benefit of citizens. In 2014, 3.4% of the agricultural land in the lowland areas of Switzerland had very high species biodiversity (BLW 2015b), this percentage was almost double in Zurich; farmers received additional biodiversity payments for the higher ecological quality on 6% of city owned agricultural land (48.6 ha). Furthermore, this percentage is also reached through advice and city specific programs such as "10,000 fruit trees for Zurich", which provides farmers with free fruit trees which they manage according to biodiversity requirements. Other examples of support payments, that city owned farms can obtain, include assistance for infrastructure costs for e.g. a farm shop or animal-welfare friendly stable constructions (pillar 3).

The city of Zurich has undertaken a range of activities to convey "Green knowledge" (pillar 4). Some "nature schools" have been installed in the city and there is a Swiss association called "School on the farm", which offers educational programs on farms in collaboration with the city department for green space management and with school teachers. The farmers participating in this program have often attended pedagogic courses, but the city farmers' interest and availability of time is often limited. As a result, the city administration has searched for other ways to spread "Green knowledge" with ideas including the provision of support to local *bottom-up* initiatives of city dwellers in smaller town quarters/neighbourhoods, who want to create accessible hobby farms (with small animals, fruit trees, berries etc.) ("Quartierbauernhöfe" e.g. Quartierhof Wynegg).

The attempts by the city to fulfil the above mentioned goals have so far concentrated on supporting the city owned farms. Urban gardening initiatives have been mainly seen by city representatives as an interesting alternative, in which "their future development will be observed". Some urban gardening initiatives however are demanding support from the city by claiming more agricultural land from city farms. A representative of the department in charge argued that this may lead to difficult situations: "If we provide agricultural land to initiatives then the farmer does not have it anymore. There are certain tensions and conflicts of interest between farmers and urban gardeners".

13.5 Land Use Approach by Zurich City Farmers

The location of farms within, or in the periphery of the city has, in most cases, resulted in diversification of the farming businesses, including direct marketing of produce; provision of multifunctional services as required by the national direct payment regulation for Swiss agriculture and additionally by the city of Zurich; and other services, such as cutting hedges for private gardeners. There is a long tradition of direct marketing through on-farm shops or direct sales to restaurants, but the interviewed farmers reported that direct marketing does not substantially contribute to their farm's revenue, but they do value the positive feedback and appreciation of their work from consumers. There is a potential to expand the marketing activities because interest in local produce by city dwellers is perceived to be growing. However, the focus of the majority of farmers is still on food production that is marketed through mainstream processing channels. One farmer commented that "farmers see themselves too much in the production function and do not fully use the cities' potential".

The strong ecological orientation of the new agricultural policy on national, cantonal and city levels is not fully supported by all groups. Implementation of nature conservation measures is mandatory on city agricultural land, but there is partly a perceived conflict in the decision between the production of food and implementation of ecological measures. Farmers deal differently with the policy goals towards multifunctional land use in Zurich. Representatives of the mainstream farming community are especially critical and are of the opinion that production of food on agricultural land should be the main goal. City policy demands may thus conflict with the recommendations of the farming community representatives. Although implementation of on-farm environmental measures has the potential to contribute significantly to farm income, farmers (in particular those on private land) are often not willing to fully implement such measures. Furthermore, farmers on private land are hard to reach through city administration support programs. Several policy actors however, mentioned an increased interest of city farmers towards biodiversity measures and explained this increase by pointing out that there are more and more projects at the municipality level that support specific biodiversity measures and thus provide more opportunities for farmers to find a way to participate. In 2015, there were six "ecological interconnectedness" projects taking place in different peripheral areas of Zurich with involvement of farmers, citizens and the city administration (Stadt Zürich 2015).

Farmers on city owned farms are obliged to open their farms to the public, such as by carrying out on-farm direct marketing activities or by hosting school excursions on their farm. Farmers, especially those running a farm shop, see it as a good opportunity to connect to consumers but they often feel that they do not have the necessary skills, knowledge or time to offer such activities. They are willing to open the farm to the public, but see it rather as the responsibility of the city to organise such activities from an organisational and educational point of view. Though, "green knowledge" in the population would also be beneficial to the farmers, who maintain cultural landscapes with recreational value for city dwellers, and so have to deal with negative aspects of public use such as irresponsible dog keepers and littering.

13.6 Land Use Approach by Zurich Civil Society

There are different groups in Zurich who cultivate (agricultural) land within the city. They range from traditional allotment gardeners with a focus on food production or a focus on the recreational value of gardening, to recent urban gardening initiatives that focus on production of food or on educational aspects. There is also an urban gardening initiative that is operated like a farming business and which is engaged in CSA.

The city has areas that have been zoned for allotment gardening. In terms of the recent urban gardening initiatives, the city department with control of green space management has only limited possibilities to provide additional land. Such land is either at the periphery of the city, difficult to reach by public transportation and preferred by farmers, or it is vacant land, where for different legal or planning reasons construction work is yet to commence. Several representatives of gardening initiatives reported that the contact with the responsible department was congenial and collaborative while they were searching for suitable locations for urban gardening. However, some initiatives reported a lack of long term commitment from public institutions, as it was not clear whether they could keep the fields and continue their activities. Investments, which are necessary if the CSA type of gardening is to be professionally managed, are hampered due to the uncertain future. Representatives of initiatives reported that they would like to be acknowledged by city policy for their contribution towards multifunctional land use but the city administration sees them rather as an experiment. One city initiative engaged in CSA activities has started to independently search for agricultural land by approaching farmers directly. The initiatives leader has observed a general interest by farmers for cooperation but, given that agricultural land is rare and under pressure in the city area, "no farmer wants to make experiments".

In 2015, the average age of allotment gardeners was around 60, although the process of rejuvenation has slowly started. Younger people are interested in allotment gardening, but many of them are primarily interested in using the garden for leisure activities. There are however young people who have an interest in gardening and producing food but who often do not want to participate in traditional gardening associations, which have strict rules and social control. The increasing demand for places for alternative forms of urban gardening has resulted in discussions within the city administration about the redesign of allotment garden areas to accommodate urban gardening initiatives and whether the allotment gardens should be partly opened to public (gaining access to the public premises). Allotment gardening associations are reluctant to accept the integration of alternative forms for several reasons, such as different philosophies between traditional allotment gardeners and gardening initiatives on how to use and cultivate the land. Furthermore, by opening the allotment premises to the public, representatives of the allotment gardening associations fear an increase in theft and vandalism.

13.7 Different Land Use Interests

The analysis shows different actor interests for land use in Zurich (Table 13.1).

Actors in policy and administration have implemented diverse support measures for the maintenance of an attractive cultural landscape, which also links to nature conservation measures and educational activities in terms of "green knowledge" for city dwellers. Food production is of minor importance for city policy and administration, but is the main focus of farmers and participants in urban gardening

Table 13.1 Different interests of actors towards multifunctional land use of urban agriculture

	Policy / Administration	Farmers	Civil society: Allotment gardeners	Civil society: Urban gardening initiatives
Recreation	+++		+++	+++
Fostering biodiversity	+++	+	+	++
Food production	+	+++	++	++
Participation and education	++	+		+++

Source: scores determined by authors based on interviews with different actors and relevant documents

+ minor importance, ++ moderate importance, +++ high importance

initiatives, although there are also initiatives in which the focus of their activities is more on the educational aspect of food production. Recreation is very important for civil society actors, although mainly for self-serving reasons rather than providing public spaces that are available for all city dwellers. Nature conservation gains a lot of support from city policy whereas the importance for other actor groups is limited. As mentioned, farmers partly see a conflict between production of food and carrying out nature conservation.

13.8 Discussion

In our study, we highlight different challenges, which the city of Zurich is currently facing regarding multifunctional land use, especially with regard to their land use policy, with different interest groups requesting land and aiming for different and sometimes conflicting objectives within the multifunctional concept. This shows potential conflicts and challenges for governance processes at the city level in order to foster sustainable territorial development and requires an integrated governance approach towards multifunctional land use concerning urban agriculture.

13.8.1 Actors' Perspectives and Challenges of Multifunctional Land Use Management

The Zurich city policy has operationalised the concept of multifunctionality into some core functions that should be provided by urban agriculture. These are expressed in the implementation of specific tools and instruments following categories in line with Bengston et al. (2004) for managing urban growth and protecting open space: (i) public ownership; (ii) regulation and (iii) incentives. Zurich is pioneering in the way in which green space management is organised and in the approaches and support schemes the city administration has in place (such as organic cultivation or support of biodiversity). This pioneering role was repeatedly referred to and acknowledged by the actors from civil society, farmers and within policy administration who were interviewed.

The aspect of food security, which is often mentioned in literature as an argument for urban agriculture in developing as well as developed countries (Barthel and Isendahl 2013), seems to be less an issue in the city of Zurich. This might explain why policy actors in Zurich put less importance on the production of food but rather on fostering biodiversity and the recreation function of agricultural landscapes. The multifunctional land use approach, that the city has chosen, is mainly framed along the lines of the multifunctional concept in Swiss national agriculture policy, however with an emphasis on education and participation of the public. This approach is in agreement with Wilson (2009) who states, that the national level often provides the framework in which policies affecting multifunctional agriculture are formulated. In this concern, one reason for framing these city policies along the federal definition of multifunctionality might be that farmers are mainly supported by agricultural funds from the national administration as instruments for rural areas are also applied for urban agriculture. Supporting farming in the city is therefore a "cheap way of land conservation" for the city administration (Grün Stadt Zürich 2006), with the city providing additional and supplementary funds. However, the policy framework at the national level, which is adequate for rural areas, might not be adequate to fully guide policy in cities. This was underpinned by Zasada (2011), who argued that the agricultural support system is rather oriented towards a continuous rural area and that spatial conditions in urban and peri-urban areas for agricultural land use are substantially different.

Local policies can have significant influence on the farming businesses. This has also been pointed out by Vandermeulen et al. (2006) who illustrated the importance of local level policies in influencing farming choices. Measures included in Zurich city policy are geared towards farmers, and especially towards city owned farms, which puts farmers in a conflict situation. Although the implementation of environmental measures can provide a significant income source, food production and nature conservation are partly seen as competitive approaches for farmers (Home et al. 2014) and not all farmers are willing to implement such measures. A greater acceptance might be gained by harmonising nature conservation and food production (Jahrl et al. 2012) and thereby supporting farmers towards "real multifunctionality" of agricultural land use.

Nevertheless, achieving multifunctional land use goals through agriculture seems to be easier to reach in the context of urban agriculture. Zasada (2011) found that farmers in peri-urban areas are often diversified in terms of marketing their products directly or diversifying on-farm activities such as farm accommodation, educational activities or health care. This diversification of farm businesses can also be observed in a lot of the farms in Zurich. Vandermeulen et al. (2006) state that the peri-urban context may not only present threats for farming businesses but also opportunities for developing new strategies.

In addition to farmers, civil society actors are increasingly becoming involved in urban agriculture in allotment gardens or in urban gardening initiatives and several scholars have highlighted their contribution to environmental, cultural and social services in cities (Middle et al. 2014; Barthel and Isendahl 2013; Plieninger et al. 2015). Garnett (2000) states that many urban gardening initiatives have clear environmental aims, such as to promote biodiversity through organic gardening practices, to reduce waste through recycling and composting and to minimise food transportation through local food production. However, green areas managed by local user groups tend to be largely neglected in urban management schemes (Colding et al. 2006). These groups are not specifically considered in strategic and long-term multifunctional land use management concepts of the Zurich city; even though civil society actors show interests that are in line with city policy goals regarding multifunctional land use (e.g. participation and education or fostering biodiversity) (Table 13.1). The organisational forms and motivations of those who grow their own food in Zurich vary greatly; ranging from interest in self-fulfilment and connection to nature, to aiming at changing the overall food-system, which would have been predicted by Renting et al. (2012). Overall, urban actors' interests for land use and in this concern different types of urban agriculture are taken into account very differently by city policy, depending on what functions it serves for public interest. This is underlined by the discrepancy in opinion between public policy and gardeners, as can be seen in the case of recreation function. Lovell (2010) states that one argument against allotment-style community gardens is that these offer more benefit to the individual gardeners than to the public, which might be a reason why policy makers/administration in Zurich have stated a goal of opening the allotment gardens to the public. Allotment gardeners predictably oppose this proposed change.

13.8.2 Actors' Competing Land Claims and Challenges of Urban Spatial Planning

Spatial planning in Zurich not only defines areas for settlements and industry, but also zones for agricultural use and hobby gardening, which do not usually compete with each other. The approach of maintaining strictly defined areas for agriculture and gardening is challenged by an increasing number of consumers who are becoming interested in food production in the city, which can be seen in the growing number of *bottom-up* organised urban gardening initiatives. Németh and Langhorst (2014) argue, that 'temporary use practices shine a light on traditional regulatory and planning systems that are based on the perceived primacy of stable and certain environments for investment as well as the avoidance of conflicting land uses'. These practices question whether traditional regulatory planning systems are adequate in terms of an increased requirement for flexibility in the development process. The rise of urban agricultural land use indicates a disparity between planning norms and standards underpinning formal land use planning processes and structures in urban development (Magigi 2008). As seen in Zurich and in other cities, many initiatives have been established on vacant land or other underutilised spaces without the permission or long-term commitment of the land owner or manager (Lovell 2010). Competing interest for land in Zurich might be explained by similar interests for land use but different forms of organisation. This is underpinned by Darly and Torre (2012) who state that "in areas where available resources are limited, the strong competition between the uses that consume these resources causes increasing conflicts and tensions." Gardening initiatives which farm more professionally, such as when they are organised as CSA, are more likely to compete with professional farmers, while less professional gardening initiatives that are organised as district or community gardens are more likely to compete with allotment gardeners. Competing demands for land may be overcome through fostering cooperation between the different actor groups, such as fostering CSAs on existing farms and integrating collective gardening in allotment garden areas. Though, the difficulty of such cooperation, might lie in the nature of their interests. For farmers, a CSA within their farm could, *inter alia*, mean an alternative income source (Brown and Miller 2008) and synergies for marketing farm products. Contrary, for allotment gardeners, community gardeners represent more serious competition for scarce land due to different approaches towards land management and the more collective and public character of gardening, which is increasingly aimed for by city policy (Lovell 2010).

13.9 Conclusions

This study has focused on the perspectives of diverse actors in shaping multifunctional land use in Zurich. Based on qualitative interviews, a workshop, and document analysis, we identified different interests for land use between city policy, famers and urban gardeners, with various actors as farmers and urban gardeners often making similar land claims. However, interdependencies between urban actor groups were found in their efforts to fulfil their diverse objectives towards multifunctional land use. These represent challenges for the actors involved; and especially for city policy and administrators. The city of Zurich holds a powerful position in governing and fostering multifunctional land use through direct control mechanisms as a large farmland owner and policy administrator. City policy has operationalised multifunctionality in terms of urban agriculture into some core aims and tasks which follow the rural development perspective of Swiss agricultural policies and mainly target farmers. However, the city sets different priorities; landscape with recreational value, biodiversity and education are more important than food production. The empirical study revealed that this approach of applying the current agricultural policies with different priority goals in the urban context is increasingly challenged, especially taking into account the multiple actor interests for land use in the city. New *bottom-up* urban gardening initiatives especially raise claims for urban food spaces, and thereby potentially address a wide range of social and cultural functions of urban agriculture. This new situation illustrates the need for an urban approach of planning and governance towards multifunctional land use in cities, which can better bridge the diverse functions of land use (e.g. combine food production with "green knowledge") and reconcile the competitive functions that are especially perceived by farmers. In such an approach, land based policies should be part of an overall concept and strategy to address agriculture and food within cities. There is a need for new and more dynamic territorial governance (Torre and Darly 2014) with a broader public and transparent debate on the multiple functions of urban agriculture and food policies in the future. It should aim at partnerships among the diverse public and private actors involved. This needs a more transdisciplinary and participatory approach that blends the views, skills, and energies of both professional and lay stakeholders (Selman 2009; Mann and Jeanneaux 2009). Further research is needed to find improved frameworks and tools that allow the different perspectives, needs and multifunctional land use functions to be reconciled and developed into a coherent land use and food city policy. The challenge will be to create suitable framework conditions, which still leave flexibility for *bottom-up* urban development in terms of new multifunctional land use approaches.

Acknowledgement The paper is based on a study which was conducted as part of the EU-funded project SUPURBFOOD ("Towards sustainable modes of urban and peri-urban food provisioning", www.supurbfood.eu) under the 7th EU-Framework for Research and Innovation. We wish to thank all interviewees and workshop participants for their time and willingness to share their expertise with us. We also wish to thank the two anonymous reviewers, Markus Wittmer from "GSZ Grün Stadt Zürich" as well as Matthias Stolze and Adrian Müller (both FiBL) for their helpful comments on this paper.

References

Aubry C, Ramamonjisoa J, Dabat MH, Rakotoarisoa J, Rakotondraibe J, Rabeharisoa L (2012) Urban agriculture and land use in cities: an approach with the multi-functionality and sustainability concepts in the case of Antananarivo (Madagascar). Land Use Policy 29:429–439

Barthel S, Isendahl C (2013) Urban gardens, agriculture, and water management: sources of resilience for long-term food security in cities. Ecol Econ 86:224–234

Bengston DN, Fletcher JO, Nelson KC (2004) Public policies for managing urban growth and protecting open space: policy instruments and lessons learned in the United States. Landsc Urban Plan 69(2):271–286

Brown C, Miller S (2008) The impact of local markets: a review of research on farmers markets and Community Supported Agriculture (CSA). Am J Agric Econ 90(5):1298–1302. https://doi.org/10.1111/j.1467-8276.2008.01220.x

Bundesamt für Landwirtschaft (BLW) (2015a) Direktzahlungen. http://www.blw.admin.ch/themen/00006/index.html?lang=de. Accessed 24 Feb 2015

Bundesamt für Landwirtschaft (BLW) (2015b) Agrarbericht. http://www.agrarbericht.ch/de. Accessed 18 Dec 2015

Bundesamt für Umwelt (BAFU) (2009) Ökologischer Ausgleich. http://www.bafu.admin.ch/landschaft/00522/01649/01650/index.html?lang=de. Accessed 03 Feb 2015

Colding J, Lundberg J, Folke C (2006) Incorporating green-area user groups in urban ecosystem management. AMBIO J Hum Environ 35(5):237–244. https://doi.org/10.1579/05-A-098R.1

Darly S, Torre A (2012) Land-use conflicts and the sharing of resources between urban and agricultural activities in the Greater Paris Region. Results based on information provided by the daily regional press. In: de Noronha Vaz T, van Leeuwen E, Nijkamp P (eds) Towns in a rural world. Ashgate, London. http://www.ekf.vsb.cz/export/sites/ekf/projekty/cs/weby/esf-0116/databaze-prispevku/clanky_ERSA_2010/ERSA2010finalpaper1325.pdf. Accessed 05 July 2015

Deelstra T, Girardet H (2000) Urban agriculture and sustainable cities. In: Bakker N, Dubbeling M, Gundel S, Sabel-Koschela U, de Zeeuw H (eds) Growing cities, growing food: urban agriculture on the policy agenda. Deutsche Stiftung für Internationale Entwicklung (DSE), Feldafing, pp 43–65

Deelstra T, Boyd D, van den Biggelaar, M (2001) Multifunctional land use. An Opportunity for promoting urban agriculture in Europe. UA Magazine 4 – Integration of UPA in urban planning, RUAF Foundation. http://www.ruaf.org/multifunctional-land-use-opportunity-promoting-urban-agriculture-europe. Accessed 25 Jan 2015

Drescher A (2001) The integration of urban agriculture into urban planning – an analysis of current status and constraints. In: ETC Urban Agriculture Programme & Swedish international development Agency (SIDA) (eds) Annotated bibliography on urban agriculture (pp 555–569). Leusden, The Netherlands. http://areeweb.polito.it/didattica/UPWARD/dwd/agricolture/dreschler.pdf. Accessed 24 Jan 2015

Garnett T (2000) Urban agriculture in London: rethinking our food economy. Growing cities, growing food. German Foundation for International Development, Feldafing, pp 477–500. http://www.ruaf.org/sites/default/files/London_1.PDF. Accessed 05 Jan 2015

Grün Stadt Zürich (GSZ) (2006) Das Grünbuch der Stadt Zürich. https://www.stadt-zuerich.ch/ted/de/index/oeffentlicher_raum/heute_und_morgen/strategie/gruenbuch.html. Accessed 15 Jan 2015

Grün Stadt Zürich (GSZ) (2015a) Landwirtschaft in der Stadt Zürich. www.stadt-zuerich.ch/landwirtschaft. Accessed 03 Feb 2015

Grün Stadt Zürich (GSZ) (2015b) Urban gardening in Zurich. http://www.fibl.org/fileadmin/documents/de/news/2015/SUPURBFOOD-Seminar_GSZ-UrbanGardening_MWittmer_Zurich_16.Jan2015.pdf. Accessed 03 Feb 2015

Home R, Balmer O, Jahrl I, Stolze M, Pfiffner L (2014) Motivations for implementation of ecological compensation areas on Swiss lowland farms. J Rural Stud 34:26–36

Jahrl I, Rudmann C, Pfiffner L, Balmer O (2012) Motivation für die Umsetzung von Ökoausgleichsmassnahmen. Agrarforschung 3(4):208–215

La Rosa D, Barbarossa L, Privitera R, Martinico F (2014) Agriculture and the city: a method for sustainable planning of new forms of agriculture in urban contexts. Land Use Policy 41:290–303

Lovell ST (2010) Multifunctional urban agriculture for sustainable land use planning in the United States. Sustainability 2:2499–2522. https://doi.org/10.3390/su2082499

Magigi W (2008) Improving urban land governance with emphasis on integrating agriculture based livelihoods in spatial land use planning practise in Tanzania. Doctoral dissertation, Faculty of Forest and Environmental Sciences, Albert-Ludwigs-Universität Freiburg, Germany. https://www.freidok.uni-freiburg.de/fedora/objects/freidok:5311/datastreams/FILE1/content. Accessed 01 Dec 2015

Mann C, Jeanneaux P (2009) Two approaches for understanding land-use conflict to improve rural planning and management. J Rural Community Dev 4(1):118–141

Mayring P (2002) Einführung in die qualitative Sozialforschung. Weinheim und Basel: Beltz Verlag

McNabb DE (2010) Research methods for political science: quantitative and qualitative approaches, 2nd edn. M. E. Sharpe, New York

Middle I, Dzidic P, Buckley A, Bennett D, Tye M, Jones R (2014) Integrating community gardens into public parks: an innovative approach for providing ecosystem services in urban areas. Urban For Urban Green 13(4):638–645

Morgan K, Sonnino R (2010) The urban foodscape: world cities and the new food equation. Camb J Regions Econ Soc 3:209–224. https://doi.org/10.1093/cjres/rsq007

Németh J, Langhorst J (2014) Rethinking urban transformation: temporary uses for vacant land. Cities 40:143–150

Paracchini ML, Pacini C, Jones MLM, Perez-Soba M (2011) An aggregation framework to link indicators associated with multifunctional land use to the stakeholder evaluation of policy options. Ecol Indic 11(1):71–80

Plieninger T, Bieling C, Fagerholm N, Byg A, Hartel T, Hurley P, López-Santiago CA, Nagabhatla N, Oteros-Rozas E, Raymond CM, van der Horst D, Huntsinger L (2015) The role of cultural ecosystem services in landscape management and planning. Curr Opin Environ Sustain 14:28–33. https://doi.org/10.1016/j.cosust.2015.02.006

Renting H, Schermer M, Rossi A (2012) Building food democracy: exploring civic food networks and newly emerging forms of food citizenship. Int J Sociol Agric Food 19(3):289–307

Schmid O, Jahrl I (2014) City region of Zurich (Switzerland). SUPURBFOOD —WP2 Final case study report. http://supurbfood.eu/scripts/document.php?id=126. Accessed 11 Feb 2015

Schmitz A, Moss C (2005) Multifunctionality in agriculture: externalities and non-traded goods. Int J Agric Resour Gov Ecol 6(3–4):327–343

Selman P (2009) Planning for landscape multifunctionality. Sustain Sci Pract Policy 5(2):45–52

Stadt Zürich (2007) Strategien Zürich 2025. http://www.stadt-zuerich.ch/content/prd/de/index/stadtentwicklung/stadt-_und_quartierentwicklung/strategien_zuerich_2025.html. Accessed 03 Feb 2015

Stadt Zürich (2013) Porträt der Stadt Zürich. Zahlen und Fakten. https://www.stadt-zuerich.ch/portal/de/index/portraet_der_stadt_zuerich/zahlen_u_fakten.html. Accessed 25 Jan 2015

Stadt Zürich (2015) Landschaftsentwicklungskonzepte (LEK). www.stadt-zuerich.ch/lek. Accessed 04 Feb 2015

Stadtrat Zürich (1999) Bericht des Stadtrates an den Gemeinderat Zürich zur Landwirtschaft. 22. Dezember 1999. GR Nr 99/635

Torre A, Darly S (2014) Land use and soils disposal: From competition to territorial governance (examples from land use conflicts in the greater Paris region). Renewable Agric Food Syst 29(03):206–217. https://doi.org/10.1017/S1742170513000379

Torre A, Traversac JB (eds) (2011) Territorial governance: local development, rural areas and agro-food systems. Springer, Berlin

Vandermeulen V, Verspecht A, Van Huylenbroeck G, Meert H, Boulanger A, van Hecke E (2006) The importance of the institutional environment on multifunctional farming systems in the peri-urban area of Brussels. Land Use Policy 23(4):486–501

von der Dunk A, Grêt-Regamey A, Dalang T, Hersperger AM (2011) Defining a typology of peri-urban land-use conflicts – a case study from Switzerland. Landsc Urban Plan 101(2):149–156

Wiggering H, Dalchow C, Glemnitz M, Helming K, Muller K, Schultz A, Stachow U, Zander P (2006) Indicators for multifunctional land use – linking socio-economic requirements with landscape potentials. Ecol Indic 6:238–249

Wilson GA (2009) The spatiality of multifunctional agriculture: a human geography perspective. Geoforum 40(2):269–280

Wiskerke JS (2009) On places lost and places regained: reflections on the alternative food geography and sustainable regional development. Int Plan Stud 14(4):369–387

Yin R (1989) Case study research. Design and methods. Sage, Newbury Park

Zasada I (2011) Multifunctional peri-urban agriculture – a review of societal demands and the provision of goods and services by farming. Land Use Policy 8(4):639–648

Index

C.-T. Soulard et al. (eds.), *Toward Sustainable Relations Between Agriculture and the City*, Urban Agriculture, https://doi.org/10.1007/978-3-319-71037-2

Kolibri 360 GmbH
Ferdinand-Jühlke-Straße 7
99095 Erfurt, Deutschland
produktsicherheit@kolibri360.de